Resource and Environmental Sciences Series

General Editors:
Sir Alan Cottrell, FRS
Professor T. R. E. Southwood, FRS

Energy Resources

J. T. McMullan, R. Morgan and
R. B. Murray

School of Physical Sciences, New University of Ulster

A HALSTED PRESS BOOK

JOHN WILEY & SONS
New York

First published 1977 by
Edward Arnold (Publishers) Ltd
London

Published in the U.S.A.
by Halsted Press, a Division
of John Wiley & Sons, Inc.
New York

Library of Congress Cataloging in Publication Data

McMullan, John T
Energy resources.

"A Halsted Press book."
1. Power resources. I. Morgan, R., joint author.
II. Murray, R. B., joint author. III. Title.
TJ163.2.M34 333.7 77-26748
ISBN 0-470-99377-4

Printed in Great Britain

Preface

The aim of this book is to present to the non-specialist a survey of energy resources and of the technologies by which they can be exploited. It is intended for students of the physical and environmental sciences in the final years at school and early years at university, as well as for that mythical beast the intelligent layman. We have avoided the use of mathematics whereever possible.

In presenting a specialised mathematical subject to a non-specialised audience in a non-mathematical manner, we are only too well aware that some degree of over-simplification is inevitable. We hope, however, that the resulting treatment is accurate.

After a short introduction, the book deals with the resources available to us, and then with techniques of energy conversion. There follow chapters on natural, fossil and nuclear sources and their associated technologies, and the book ends with a chapter on waste. We hope that the reader will find most of what he wants within these pages, but for those who wish to know more, there is a bibliography of further reading. We must warn the reader, however, that he will *not* find here a prophecy of doom, nor will he find an answer to the world's energy supply mysteriously conjured up out of burning camel dung. Reality is not like that. The world will not come to an end just because other people leave lights on; on the other hand there will undoubtedly be a considerable problem over fuel supplies during the next few decades, which no single technology and no single resource can easily overcome. If the book succeeds in convincing the reader of the truth of these assertions, it will have achieved its purpose.

In writing this book, we have drawn on a wide range of sources of information, and although specific acknowledgements are made in the text it is impossible to acknowledge all of them; we would like to record here our thanks to all those who have so willingly offered us advice and information.

Coleraine
1976

J. T. McMullan
R. Morgan
R. B. Murray

Contents

1 Introduction

The initial panic reaction to the 'energy crisis' has now subsided and people have begun to realise that what we really face is a chronic situation. Energy supplies, or, more exactly, fuel supplies, are limited, and for the most part exhaustible. Further, the technologies for handling the extremely long term sources are very poorly developed. This has led to an upsurge in the number of forecasters and forecasts. These forecasts range from the sublime to the ridiculous and from the rose-tinted to the cataclysmic. It has become fashionable to take the fact that oil reserves are known to be limited and to have a lifetime of 25 years (as estimated in 1973 and now reckoned to be too low) and to prophesy that after A.D. 2000 there will be *no more fuel,* that *electricity generation will stop,* that *vehicular transport will cease* and that *life as we know it will be at an end.* These assertions are patently absurd. It is certainly true that oil reserves are finite, but they are not going to dry up abruptly on 31 December 1999. Instead, oil will become more expensive and there will be a natural diversification to other sources of energy brought about by this price barrier. Meanwhile there will certainly be a move to re-establish coal as the dominant fossil fuel since there is considerably more coal in reserve than there is oil. This will be discussed in Chapter 2. Thirdly, nuclear electricity generation will be extended, though once again this is not a panacea and will only provide any long term benefit if breeder reactors become commonplace, or, preferably, if thermonuclear fusion can be made to work. Even after these steps have been taken, it is still true that fossil reserves are finite, and it is probable that their proper place is as feedstock for the chemical industries rather than as fuel for fires. The only possibilities of conserving them for these purposes lie in reducing waste and in substituting other sources of energy over an extended period.

How has the so-called crisis arisen, and what are the alternatives open to us for coping with it? Further, what are the prospects for the long term supplies of fuel and energy? These are the questions that we hope to answer in this book. The background can be given in this introduction, but the other questions require much more detailed treatment and will be answered in the subsequent chapters. Further, because this book is one of a series, certain aspects of energy production, primarily pollution, will be given only a limited treatment as they will figure in greater detail in the other volumes.

What is the background to our present energy problem? The warnings that it would arise are not new; they have been voiced for well over seventy-five years, and indeed in the middle nineteen fifties, before the enormous increase in oil consumption of the late nineteen fifties and the nineteen sixties, clear and authoritative data were being presented within the oil industry by geologists such as M. King Hubbert that the year 2000 would see the end of the 'oil age'. These data were based on known rates of discovery of new oil fields, and on known and extrapolated figures for oil consumption. As such they were open to the usual criticism that the basic assumptions were imprecise but, nevertheless, they clearly indicated the nature of the problem. Sooner or later, the rate of consumption of petroleum products would exceed the rate of discovery of new reserves. Once this stage was reached, the end of oil as a fuel would be in sight. The situation with oil is further complicated because of the geological characteristics required for its accumulation and also because of the rise of the petrochemical industry. As will be discussed later, in Chapter 5, oil and coal are formed by similar processes of sedimentation and compression. However, coal is a solid material and when formed remains in place. Oil and gas, on the other hand, are fluids and will migrate and escape unless the overlying rock strata are impermeable. This creates a more stringent geological limitation on the occurrence of oil and gas deposits. As a consequence coal is common throughout the world but oil and gas, while widely distributed over the earth's surface, occur in isolated patches where suitable sedimentation beds were later overlaid with suitable impermeable caps. Thus, oil and gas reserves will be smaller than those of coal, and, incidentally, will be more difficult to locate.

The petrochemical industry adds a new dimension to the problem of fossil fuel use. In the post-war period there has grown up a massive industry producing synthetic fibres, pharmaceuticals, fertilisers, plastics and many other products. Without these life would be very different from what it is today. This industry is based on the use of oil as chemical feedstock and so has contributed greatly to the increase in consumption throughout this period. However, the exhausting of oil reserves would create problems within this industry that would be only partially solved by a changeover to coal as the basic raw material. This aspect adds an important factor that must be considered in great detail in planning for the use of fossil fuels in the coming decades. In the words of Sheikh Yamani, 'Oil is too precious to burn'.

This is not the first time there has been a fuel crisis. As long ago as the seventeenth century there was an acute shortage of fuel wood and charcoal in England and this led to shortages of iron over a period of about one hundred years. This, coupled with the invention of the steam engine in the mid-seventeenth century and its later development by Newcomen in 1712, led to the introduction of coal as a major fuel and the replacement of charcoal by coke in the smelting process. By the nineteenth century well over 90% of Britain's fuel supplies were derived from coal. Coal formed the basic fuel for the industrial revolution and remained the dominant fuel in Britain until well after the second world war when the comparative cheapness of oil and its greater convenience led to expansion of the use of oil as a fuel and relative contraction of the use of coal. In fact, in Britain, oil did not become the dominant fuel source until the early nineteen seventies, when it was becoming ironically apparent that oil reserves were limited.

By comparison with coal and oil the other sources of energy in Britain are very small. Nuclear power in 1974 produced only 29 396 GWh of electricity out of a total output of 250 484 GWh, that is, 11.74%, while in the period January–May 1975 the corresponding figures were 11 960 GWh, 114 762 GWh and 10.4% (Energy Trends, 1975). Hydroelectricity formed an even smaller fraction of the total output at approximately 2%.

The changing pattern of fuel use in Britain is clear from Fig. 1.1 which shows the percentage consumption of different fuels in Britain over the period from 1850 to 1974. It is important to remember that these are percentage data and not absolute values. They merely indicate the relative importance of the various energy sources. Fig. 1.1 clearly shows the early industrialisation of the British economy and also, through the tenacious hold of coal over the energy supply, the inertia of the industrial system and reluctance to change over to the newer oil-fired plant for a considerable part of the twentieth century. By contrast, Fig. 1.2 shows the corresponding data for the United States of America. Important similarities and disparities between the two diagrams are apparent. There is the same general

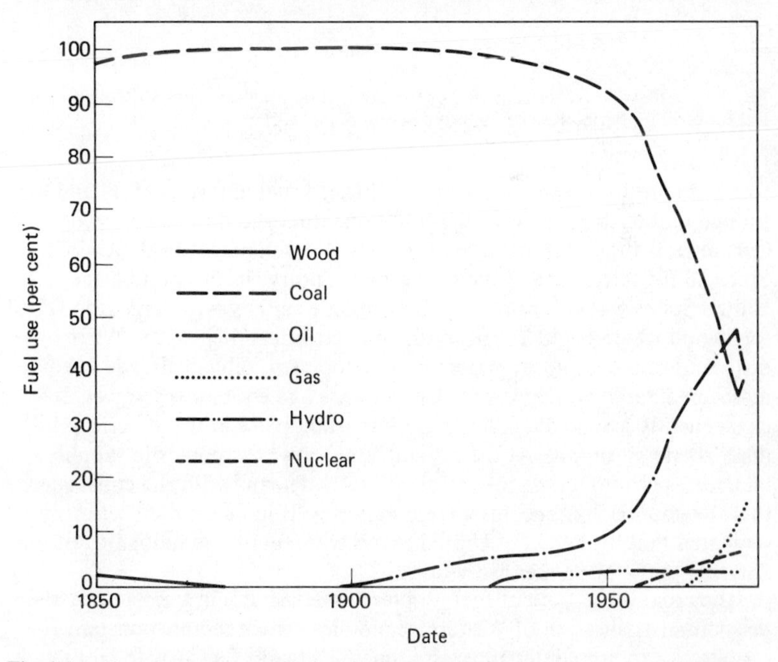

Fig. 1.1 Changing patterns of fuel use in the United Kingdom. (From J. T. McMullan R. Morgan and R. B. Murray, *Energy Resources and Supply*, Wiley 1976. Reproduced by permission of John Wiley and Sons Ltd.)

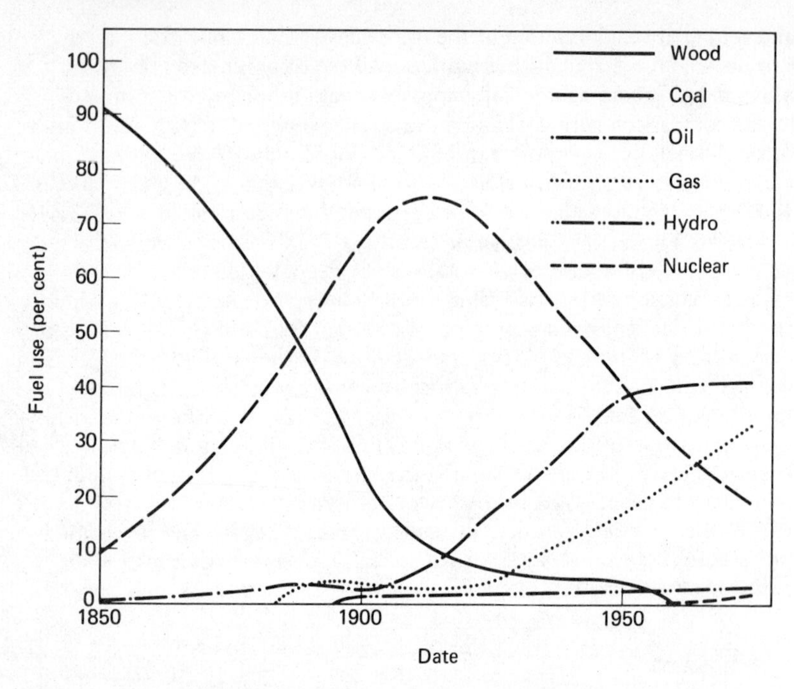

Fig. 1.2 Changing patterns of fuel use in the United States. (From McMullan et al. Reproduced by permission of John Wiley and Sons Ltd.)

trend of a decline in the use of the traditional fuel in favour of oil and gas, though in the American case the traditional fuel was not coal but wood. This reflects the different state of industrial development in the United States in the latter part of the nineteenth century. In this period, the United States was still primarily an agrarian economy with large reserves of fuel wood which could be cut by the homesteaders themselves. When industrialisation became an important development, oil had already been discovered and it was apparent that America had enormous reserves. It consequently assumed a much more important place in the American industrial energy supply. At the present time, however, domestic supplies of oil and gas can no longer meet United States demand with the consequence that the country has become a large importer. It has even been seriously suggested that by 1985 the United States will require the unloading of one 250 000 tonne oil tanker every hour.

The crux of the present fuel problem therefore lies in two factors: the heavy industrialisation of Western economies, which require constant fuel supplies for their maintenance, and the unavoidable fact that fossil fuel reserves are finite and exhaustible. What can be done about this situation? One school of thought maintains that we should lower our expectations and return to a much simpler agrarian economy. Unfortunately this is probably not possible in the short and medium terms at least. Because of industrialisation, society has changed from one in which the population is largely distributed over the land to one in which most people live in large conurbations and the food supplies are produced in more efficient units employing fewer people. In addition the population has grown from that which could be supported on the older system to the much higher level that can be supported on the new system. To return rapidly to the old system would cause considerable distress. Further, the land used to be cultivated by means of work animals. These are much vaunted in some circles for their good qualities of low fossil fuel consumption and recyclable waste production. They are also cuddly when young. Unfortunately again, there is a difficulty. They must be fed and, in order to feed them, land must be devoted to growing food for them, and them alone. In 1918, there were 25 million work animals in the United States, which by then had most of its arable land under cultivation. One quarter of the total harvested crop was required as feed for these animals. By contrast, tractors do not eat grass. Finally, present population levels and city structures require highly sophisticated communications systems both for the supply of goods and for the maintenance of organisational cohesion. The maintenance of these systems requires a ready supply of suitable fuels.

This then is the background to the problem. Western society is highly mechanised and has a high expectation in living standards. Automatic equipment has replaced manual effort even in the home, and, in addition, we have become heavily dependent on one particular fuel – oil. This is proving to be of extremely limited duration at present levels of consumption and also has proved to be highly susceptible to political factors because certain fairly small regions of the world dominate its supply.

There are alternatives to oil and gas. Coal is the first obvious choice. Because of the nature of the formation processes, coal is widely distributed throughout the earth's crust and large deposits lie within the territorial boundaries of most industrial countries. It is, however, less convenient to use than oil and gas, and is also more expensive. (This is true in real terms as the present high cost of oil is not a reflection of the real cost of extraction, but is rather a reflection of what the producing countries feel should be paid for it. North Sea oil, by contrast, is an expensive fuel because the extraction costs are higher than in the Middle East.) Coal is also fairly labour intensive and in general raises further problems of waste disposal. However, the deposits are very large and, as will be shown in Chapter 2, are sufficient for several hundred years.

Nuclear power is also widely considered as the best solution to our energy difficulties. Once again, there are problems. Known world reserves of uranium at economical mining costs are about 1 million tonnes while projected requirements to 1990 are 1.8 million tonnes. (IIASA Conference Report, 1975). There is, therefore, an impending shortage if present trends continue and if the present so-called burner reactors remain the only type of nuclear power generating plant. Fortunately, however, there is a solution. Part of the process involved in every nuclear reactor involves the conversion of some of the useless uranium–238 to fissile plutonium–239. This normally happens only on a small scale, but, in the so-called *fast reactor,* it is possible to encourage the reaction to such an extent that more fuel is produced than is consumed in the power producing fission processes. This is the principle of the *fast breeder reactor.* If these reactors were to become commonplace, the useful lifetime of uranium reserves would be extended by several hundred years.

The other major difficulty with nuclear power has already gained much publicity. It is undoubtedly true that radioactive wastes present a long term hazard that is greater than anything with which man has had to cope before. The whole philosophy of the containment of these waste products assumes that present civilisation is sufficiently stable that after three, four or five hundred years, records will still remain, and be understood, that explain the reasons for the existence of the storage farms, the nature of their contents and the ways of handling them. In the light of previous experience this is perhaps a little presumptious. A more immediate problem is the possible result of an accident in the operation of a reactor. Critics of the nuclear power programmes insist that the risks accruing from such an accident are so great that they should not be taken. On the other hand, it must be pointed out that none of the fearful situations predicted have yet occurred and that the nuclear power industry has an extremely good accident record, better than other industries. The precautions against accidents and the defensive steps once an accident has occurred are painstaking and thorough. Previous experience and theoretical models of reactors are used to predict the safety systems to be adopted. Nonetheless, the basic criticism is still valid. It is absolutely true that no matter what steps are taken to circumvent accidents they will still occur, if only because an accident, by definition, is unforeseeable. Nuclear power and the problems of radioactivity and reactor safety will be discussed in Chapter 6.

Coal, oil, gas and nuclear sources together with hydroelectricity, are the main claimants to the title of major energy producer (Fig. 1.3). There are, of course, others, but they are all of rather less utility on a global scale. This is true for a variety of reasons. The main group of such alternatives is that of the so-called *natural energy sources.* These are solar energy, geothermal energy, wind power, tidal power, wave power, hydroelectric power and power extracted from the oceans by making use of the temperature differences that exist both horizontally (by using ocean currents) and vertically. All of the natural energy sources have the advantages that they are available without any further effort on man's part, that they are free and that they are extremely long term. However, the disadvantages are large and various. The technologies for their use are poorly developed and likely to be expensive. Sometimes, as in the case of solar power and tidal power, they are not

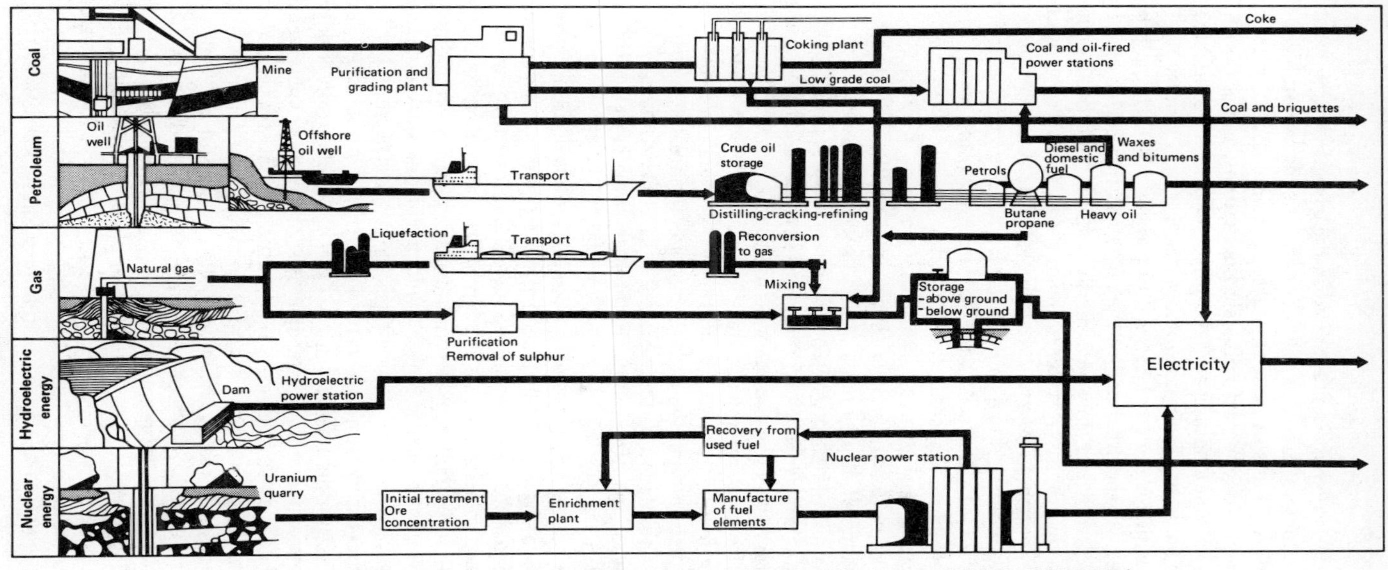

Fig. 1.3 Major energy sources and their interconversion (modified from UNESCO COURIER).

continuously available. Sometimes, as in the case of wind power, there is both the problem of lack of availability and the inverse problem of an unwanted surfeit. Thus storage is an inherent requirement of most natural energy schemes, to cope with possible fluctuations, and also with the unavoidable fact that a natural energy source need not be available at the most convenient moment. The most obvious case for such storage is the mill pond or dam by which water can be stored to provide power at the precise time it is required. At the other end of the energy storage scale is the concept of the *hydrogen economy*, in which energy is stored by producing hydrogen, possibly by electrolysis of water in the case of there being surplus electricity available, or possibly by some other process. When the energy stored in the hydrogen is required, it can be burnt to produce heat and water once again, or used as fuel for a fuel cell to produce electricity and water. Incidentally, the water produced is absolutely pure and so the hydrogen economy also produces a ready supply of good drinking water!

The use of our natural power sources is not without its drawbacks. For example, windmills can be noisy and hydroelectric dams can silt up (and once silted up it is difficult to see how they can be repaired). Hydroelectric schemes can also have a deleterious effect on their local environment. For example, the Aswan dam, which was supposed to serve the dual purpose of supplying electricity and controlling floods in the Nile Valley, has had the secondary effect of trapping the nutrients that were normally swept downstream and deposited in the delta. Consequently, for the first time farmers in the Nile Valley have had to use artificial fertilisers. Meanwhile, they have not been swept away by flood as in the past, so perhaps this is worth a little fertiliser.

All solar energy schemes suffer from the serious disadvantage that the sun's energy arrives on the surface of the earth with a fairly low energy density, about 600 watts per square metre. Consequently, any solar power station of commercial size must occupy a large area. Also, once again storage is an important consideration because cloud cover and night time seriously affect the level of power production from this source. Solar energy does have several important advantages, however. The amount of energy that comes to the earth from the sun is enormous, over 10 000 times man's present consumption. It is free, and its use is completely non-polluting. Much of the necessary technology already exists; much is a trivial extension of that which does exist. It is really only in the field of solar electricity production that significant progress must be made before engineering applications can become a reality. Even here, the devices and materials to generate and exploit solar electricity do exist. The problem is that they are relatively inefficient and, even more important, they are extremely expensive both in money and in energy terms. Thus, power plant based on electricity production from solar energy involves a capital cost about twenty times greater than the corresponding facility based on fossil fuel boilers.

This is the nub of the problem in implementing natural energy sources –

expense. The capital cost of implementation is almost always high – much higher than that of conventional sources. It must be remembered that wind and water mills used to be commonplace and provided much prime motive power for industrial society. (In fact, in the north of Ireland, many former linen mills still have working water wheels or turbines now producing 30 kW of electricity and more). The only reason these facilities died out was because the 'new' electricity and oil fuels provided cheaper, more convenient and more flexible power sources. They will not return to common use until either their economics improve or the other sources become so expensive that alternatives must be found. Undoubtedly the fuel price escalation resulting from the increased price of oil will provide a spur to their implementation.

The other long term hope for a solution to the energy problem is nuclear fusion. In fusion very small nuclei are forced together to fuse into one larger one with the emission of energy. In the short term fusion power depends on the reaction between deuterium and tritium, both heavy forms of hydrogen. Deuterium is extremely common in the sea, but tritium, being radioactive with a fairly short half-life, does not exist in nature in any great abundance. Consequently, it must be manufactured from an isotope of lithium through nuclear reactions. The lifetime of this form of fusion power therefore depends on world reserves of lithium. These are probably equivalent to coal in energy terms. Much more attractive, however, in the long term, is the possibility of the much more difficult reaction between two deuterium nuclei. This is considerably more difficult to achieve, but the rewards of so doing are enormous. Reserves of fuel would become available that would allow us to increase our consumption of energy by a factor of 1000 in global terms, and the reserves would then last for over one million years. This would really be the utopia in power generation, but unfortunately this situation is a long way off. Fusion has not yet been seen to work in anything like the correct conditions for the controlled production of power, but it is fair to say that much progress is being made and that the likelihood of success is now seen as being much greater than it was in the late nineteen sixties.

The other aspect of energy production and use is conservation. Once it is fully realised that we are dealing with a finite resource it becomes important to take steps to conserve what we have. Thus various governments are encouraging the installation of improved insulation in buildings and are introducing speed limits to conserve motor vehicle fuel. Various possibilities present themselves. Fibreglass, polystyrene, polyurethane, urea formaldehyde, and mineral wool are all common insulating materials. The use of any of these definitely results in a reduced fuel consumption in space heating applications and is therefore a GOOD THING. The question that is never asked, however, is the extremely important one – how long is the pay-back period in *energy terms*? That is, how long does it take to recoup the energy expended in producing and installing the insulation? Also, since the plastic insulating materials are petrochemical in origin, how much oil is actually conserved by their use? This type of consideration leads to the area of *Energy Accounting* which attempts to determine the cost of goods in terms of the energy needed to produce them, instead of money value. In an ideal world, the two would probably be fairly closely related, but, at present, there is no real connection between energy cost and money cost. Energy accounting is important in the long term as the only way of finding the true energy cost of various measures that are undertaken. For example, it is well known that double glazing a window saves energy. However, the money cost of doing so is such that the time taken to recoup the capital cost in fuel savings is about fifteen years – a marginal advantage. What is the time taken to recoup the energy cost of producing the glass, transporting it from the glassworks to the house and installing it? In view of the high energy cost of glass it is likely to be about the same as for recouping the money cost. This raises the entire question of whether or not a given conservation proposal is worth pursuing in terms of the return in energy.

Meanwhile the major problems of energy conservation remain. On the domestic scale there is little that can be done other than to insulate, reduce temperatures somewhat, install heat pumps (which have a primary fuel efficiency of better than 100% as shown in Chapter 7), and hope that the oil does not give out too soon. On the industrial scale, however, there is considerable scope for improvement. Much of industry produces vast quantities of waste heat. Much of this waste heat is *low grade*, that is, it is at such a low temperature that it is not of any real use to the industrial process, but is of sufficiently high temperature that it is embarrassing to throw out as effluent. It is important that this effluent be used to preheat incoming water in process steam generation plant, or to preheat the feed to heating systems or some of a series of similar possible applications. In addition, industrial architecture leaves much to be desired from an energy conservation viewpoint. Once again, glass office blocks and buildings with large amounts of glass (either single or double glazed) cause enormous losses in energy and waste of fuel for two reasons. First, when the sun is not shining, they constitute high loss regions for the building so that fuel bills are high during the winter. By contrast, they act as greenhouses when the sun does shine, and, for industrial applications, must be ventilated heavily or air-conditioned. This is not as extravagant as it sounds, because it is important to keep temperatures in working environments reasonably comfortable, otherwise productivity decreases. The crime of glass buildings is that they combine the bad features of low thermal insulation and very short time constant. Consequently, they heat and cool rapidly in response to the appearance and disappearance of the sun. They should be compared with the thick-walled buildings of desert countries. These have heavy insulating properties, but more important in many ways, they have a long time constant. This means that their response is slow, and, in fact, these houses reach their peak temperature in the late evening, and are at their coolest in the middle of the day. There is no reason why similar 'good' design could not be applied in other climates with corresponding improvement in fuel economy.

Finally, what are the lifetimes of the world's reserves of various fuel and

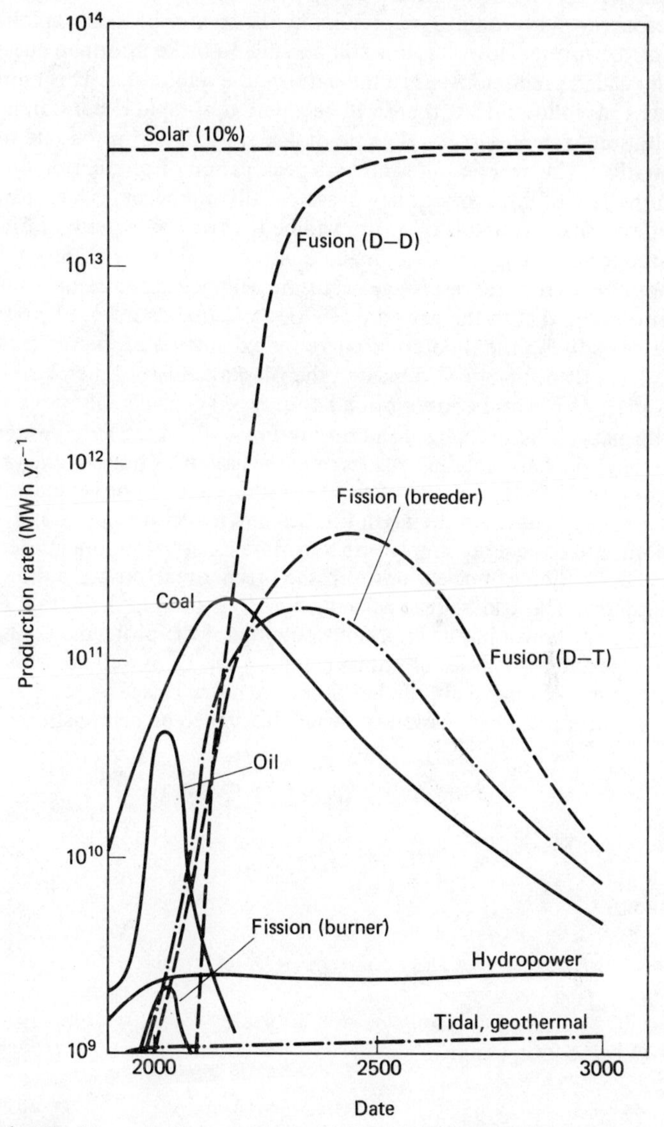

Fig. 1.4 Possible projections of resource lifetimes. (From McMullan et al. Reproduced by permission of John Wiley and Sons Ltd.)

energy supplies? It is possible to estimate the lifetime of a finite reserve in a fairly straightforward manner if we know two things: firstly, the magnitude of the reserve, and secondly, the pattern of exploitation. Armed with these one can predict exactly how long a reserve will last. In practice, it is more difficult. Firstly, we do not know exactly the magnitude of our reserves of any resource, and secondly, we only know the history of their exploitation and not the future. However, it is still possible to make informed guesses (usually called predictions) as to the pattern of exploitation. This must be somewhat as follows. First there will be a period of rapid expansion in the exploitation of the reserve as the rate of discovery outstrips the rate of exploitation. This will be followed by a peak period of production during which the rate of discovery of new reserves falls and becomes less than the rate of exploitation, and this in turn will be followed by a period of falling consumption as the reserves are finally consumed. Thus the graph of the consumption of a finite reserve against time will look like a camel's hump. It is interesting that in the period to 1970, the actual data for oil consumption lay exactly on the theoretical curve for exhaustion of the reserves by 1995. Since then there have appeared the Alaska fields and the North Sea discoveries. Also world consumption has dropped considerably since 1974. Thus the patterns have changed and the lifetime of the reserves has extended.

We can make fairly intelligent guesses at the usable lifetimes of the various resources based on the known levels of consumption and projected extent of the reserves. These are shown in Fig. 1.4 and it is clear that oil is a small contributor to our energy supply on a long time scale. Even more interesting is the contribution of nuclear power if the breeder reactors are not implemented rapidly. This can be seen as being literally a drop in the ocean. By contrast, solar power has an enormous potential if the problems of high capital cost and difficulties of implementation can be overcome, while nuclear fusion is potentially the best hope we have. This assumes that it does become a practical possibility, which has yet to be demonstrated.

2 Energy Resources

Solar power, in the form of radiation from the sun, is the ultimate source of almost all the world's energy, including coal, oil, natural gas, wood, food, wind and waves, as well as direct heating by the incoming radiation. It is not a truly eternal source; all astronomical evidence indicates that the sun will eventually suffer a catastrophic end. However, it is all but certain that it will last rather longer than mankind, so we can regard it as eternal without any serious error. Furthermore, the rate at which solar power reaches the earth is fairly constant, and will probably remain so for the next several millenia. There is little that we can do to influence it, which is perhaps fortunate! Solar power can be regarded, therefore, as a truly inexhaustible supply.

Solar energy manifests itself at the earth's surface in a number of different forms. The most obvious one is *direct* radiation, that is, the type of radiation which forms shadows, and known colloquially as sunshine. Equally important is *diffuse* radiation which arises when the direct radiation is scattered by clouds and other atmospheric artifacts. Both of these forms of radiation can in principle be utilised directly by man, but because direct radiation can be focused, while diffuse radiation cannot, the technology for harnessing it, and the efficiency likely to be achieved in doing so, differs in the two cases (see Chapter 4). Solar energy causes evaporation of water, and hence rainfall, rivers, and so on. This is used as a source of power for hydroelectricity. Solar heating of the earth's surface gives rise to atmospheric circulation, and hence winds, and the effect of wind on large bodies of water is to generate waves. Both wind and waves are potential sources of useful power, although their exploitation is by no means straightforward. Solar heating also generates ocean currents and temperature gradients which have been suggested as a source of power, although there are objections, both technical and environmental, to their use.

Green plants are able to utilise solar energy to convert atmospheric carbon dioxide and water into carbohydrates and other organic molecules. These form foods for animals (including man) and a wide range of useful materials, many of which (e.g., wood) can in principle be used as fuel. However, there are more fruitful ends to which they can be put, and their use as fuel should be regarded as rather wasteful. Green plants are also the original source of fossil fuels: coal, oil, peat and natural gas. There is evidence to suggest that fossil fuels are still being formed, although considerably more slowly than they are being used, so that even though they are not truly exhaustible, they behave to some extent as though they are. This is considered in more detail in Chapter 5.

All the above are manifestations of solar power. In addition, there are th non-solar sources. Tidal energy arises from the gravitational attraction between the earth and the moon (and, to a much lesser extent, between the earth and the sun). In contrast to solar energy this is not an inexhaustible source of power, because any energy extracted comes from the rotational energy of the earth; in other words, the earth slows down and the day gets longer. The effect of any practicable rate of power extraction on the length of the day is, however, likely to be small compared with the existing natural dissipation. Nuclear power, in the form of reserves of uranium and deuterium and other possible nuclear fuels, is another non-solar source, and is again exhaustible, although as we shall see in Chapter 6 the amount of power available from this source is potentially very large. Finally, geothermal heat, which is probably partially attributable to natural nuclear decay of radioactive materials in the earth's core, is again a non-solar source, and is again exhaustible in principle, although because the rate at which it can be exploited is rather limited at present (see Chapter 4), it has many of the features of a rather small inexhaustible supply.

The quantities of power involved in the inflows of energy from solar, tidal and terrestrial sources have been estimated with fair accuracies and are illustrated schematically in Fig. 2.1. In assessing the scale of the flows involved, it should be borne in mind that the current world total of

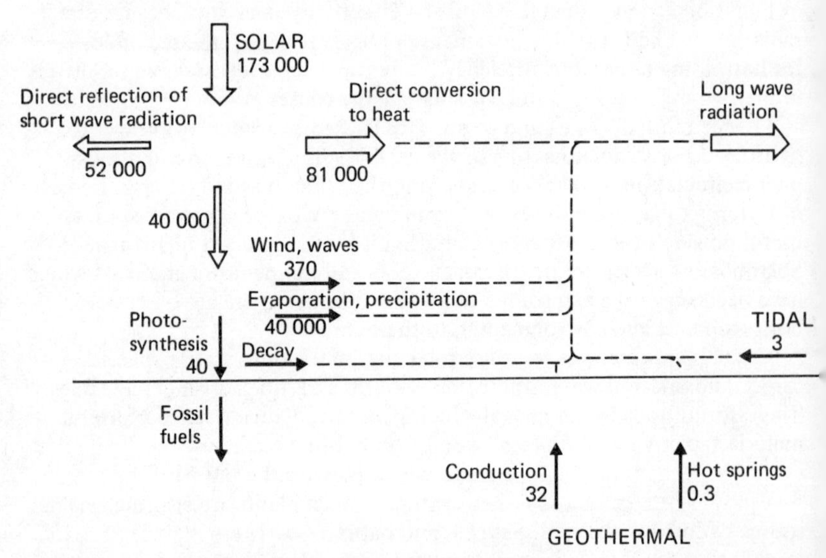

Fig. 2.1 The earth's energy budget. A schematic representation of the energy flows to and from the earth's surface. The units are terawatts.

'artificial' power used by man is about 6 x 10^{12} watts, which is almost negligible in comparison with the huge natural energy flows. Thus it can be seen that the so-called energy crisis is no such thing!

We shall now turn to a detailed study of the supplies of energy available from each of these sources and, given the central importance of solar energy, it is natural that we should begin by analysing the solar radiation that reaches us after its eight minute journey through the void of space.

The spectrum of the electromagnetic radiation emitted by the sun extends from wavelengths of less than 1 nm (10^{-9} m) to many hundreds of metres. This ranges from high energy X-rays to low energy radio waves but centres on the ulta-violet, infra-red and visible regions (see the classification in Fig. 2.2). In fact, some 98% of the total emitted radiation is carried

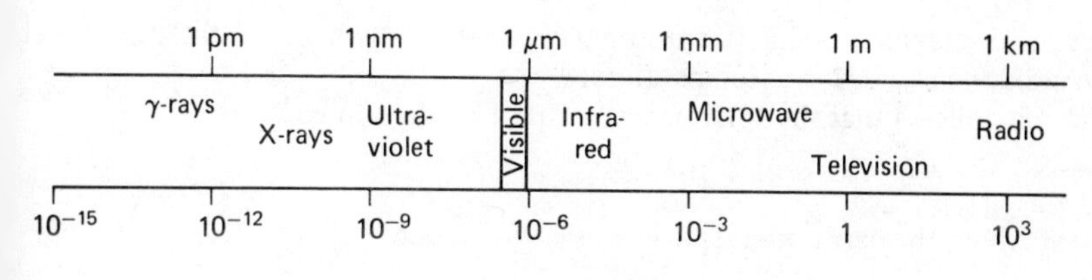

Fig. 2.2 The electromagnetic spectrum.

by wavelengths between 250 and 3000 nm. Between these limits the sun is effectively a 'heat radiator', that is, a body which emits radiant energy by virtue of its temperature. Indeed the sun may be compared to a 'black body', a hypothetical entity which completely absorbs all the radiation incident upon it whatever the wavelength. The radiation emitted by such a body depends only on the temperature of the body and on no other physical parameters – hence the usefulness of the concept for the scientist.

The spectral intensity (energy flux per unit wavelength) of black body radiation is well known and can be described adequately by a number of theoretical formulae. (Originally, of course, the study of black body radiation was instrumental in establishing the concepts of quantum physics.) If we consider the sun to be a uniformly radiating body and compare its spectral curve (as recorded from above the earth's atmosphere by rocket borne instruments) with black body distributions for various temperatures, then we find that the solar curve approximates well to that for a black body at 6000 K.

The solar constant, S, is defined as the amount of energy per unit time passing through unit area at right angles to the direction of the solar beam, measured just outside the earth's atmosphere. By treating the sun as a black body of radius R_S and temperature T_S we can calculate the solar constant. Stefan's Law tells us that the radiation emitted per unit surface area is σT^4, where σ has the value 5.67×10^{-8} W m^{-2} K^{-4}. Thus the total power emitted by the sun is $4\pi\sigma R_S^2 T_S^4$. By the time this radiation has travelled a distance R outwards from the sun it has spread out over an area of $4\pi R^2$, so that the power transmitted through unit area is given by

$$S = \left(\frac{R_S^2}{R^2}\right) T_S^4$$

Astronomers tell us that R_S is 7×10^5 km and the earth–sun distance, R, is 1.5×10^8 km. This leads to a value for the solar constant of 1600 W m^{-2} which is in reasonable agreement with the currently accepted experimental value obtained by measurements from rockets above the atmosphere:-

$$S = 1360 \text{ W m}^{-2}.$$

The discrepancy is due to approximating the sun by a disc of fixed diameter radiating at a uniform temperature.

It follows that the total power emitted by the sun is

$$4\pi R^2 S = 3.85 \times 10^{23} \text{ kW}$$

of which the earth intercepts energy at the rate of

$$\pi R_E^2 S = 1.8 \times 10^{14} \text{ kW},$$

a truly staggering figure. By comparison the present total world consumption of power is close to 10^{10} kW, over ten thousand times less than the input of solar power. But, of course, this power is needed not at the top of the atmosphere but at the earth's surface (unless it is intercepted by solar panels in orbit) and, before the radiation reaches ground level, it must run the twin gauntlets of atmospheric absorption and reflection.

A considerable proportion of the radiation emitted by the sun lies in the ultra-violet region of the spectrum. Ultra-violet radiation is extremely harmful for living systems – try looking at the discharge from a mercury vapour lamp for some time and you will have very sore eyes! This is because the energy carried by a photon of ultra-violet light (4 eV or more) is larger than typical chemical bond energies and can therefore bring about the destruction of the molecules which are essential to life processes. Fortunately our atmosphere has developed in such a way as to shield us from this unwelcome component.

Biologists tell us that the primordial atmosphere of the earth contained no free oxygen and was virtually transparent to ultra-violet radiation. Consequently, life could only exist in the sea. Once micro-organisms in the oceans developed the photosynthetic process (see later) oxygen was released into the atmosphere and, because the metabolic processes of primitive organisms were primarily anaerobic, this oxygen content was able to build up. As oxygen molecules diffused upwards they were themselves decomposed into free oxygen atoms, some of which then combined to form molecules of ozone, O_3. Ozone has the property of strongly absorbing ultra-violet light and so, once established in the higher reaches of the earth's atmosphere, it could act as a stratospheric filter for these harmful rays. Now photosynthetic plants could grow on land and influence the development of the atmosphere to its present composition.

The atmosphere has a second essential role to play in maintaining conditions on the earth's surface which are hospitable to life as we know it. To illustrate this we shall calculate the effective temperature T_e of the earth based on a radiative equilibrium between incoming solar radiation and outgoing 'earth energy' (assuming that the earth radiates as a black body at a temperature of T_e). The amount of solar radiation intercepted by the earth is equal to the solar constant multiplied by the earth's cross-sectional area, πR_E^2. To allow for the 'albedo' or reflectivity of the earth's surface, denoted by A, we must include a factor of $(1 - A)$, giving us an incoming energy flux of $\pi R_E^2 S(1 - A)$. In equilibrium this must equal the outgoing flux:

$$4\pi R_E^2 \sigma T_e^4 = \pi R_E^2 S(1 - A)$$

$$T_e = \left[\frac{(1 - A)S}{4\sigma}\right]^{\frac{1}{4}}$$

Taking A to be $\frac{1}{3}$ on average, we arrive at a value for T_e of 253 K or −20°C. But we know that the surface temperature of the earth is much greater than this (approximately 13°C on average), so where has our calculation gone wrong? It hasn't! We have in fact calculated the effective temperature for the composite earth–atmosphere system. The radiation emitted from the earth's surface is mainly in the infra-red and this is just the region in which molecules such as carbon dioxide and water are strongly absorbing. The radiation which they absorb is then re-radiated in all directions so that some of it is directed back towards the surface. Consequently, the earth's surface remains warmer than we have predicted – this is termed the 'greenhouse effect' for obvious reasons.

Thus in addition to protecting the earth's surface from harmful ultra-violet radiation the atmosphere also acts as an insulating blanket. Of course, we can never get something for nothing and in return for the atmosphere's protective role we must accept that it also prevents much of the sun's energy from reaching the ground. We have already alluded to the absorption of light by gases such as carbon dioxide, water vapour and ozone, and we must now consider the effects of atmospheric scattering and absorption.

Scattering occurs when small particles in the atmosphere redirect the solar beam. If we assume that we are dealing with spherical particles which are small in comparison to the particular wavelength of light, then the phenomenon of Rayleigh scattering is observed. It is possible to show theoretically that the amount of radiation scattered by the particles is inversely proportional to the fourth power of the wavelength, that is, it varies as $1/\lambda^4$. This means that light of short wavelength (blue) is much more effectively scattered than light of long wavelength (red) This is just what we should expect intuitively – a small particle will interact most strongly with light of a similar wavelength.

We all know the effect that this has on the sunlight which we observe at the ground. In a 'pure Rayleigh atmosphere' in which only air molecules are responsible for scattering, the sky appears to have a blue colouration as light in the blue region of the spectrum is more easily scattered than the red. As the path length of the sun's rays through the atmosphere increases, a greater fraction of the blue component of solar radiation is scattered by the greater number of molecules along the path. Consequently, the sun appears to become redder while the sky appears more blue. The effect is most marked, of course, when the sun is very low in the sky, as at sunset. However, in the presence of larger particles, such as water droplets or ice particles, light of all colours is scattered to a greater extent with the result that the sky appears less blue and will eventually appear to be white when a sufficient number of large particles is present, as in a cloud.

Whereas scattering takes place at all wavelengths, absorption is a selective process. The molecules in the atmosphere which are mainly responsible for absorption of solar radiation are H_2O, CO_2, N_2, N_2O, O_2, O_3, NO, CH_4 and CO, while free radicals of oxygen and nitrogen also have a part to play. The visible region is relatively free from absorption bands whereas substantial absorption occurs in the ultra-violet (O_2, N_2, O_3, N, O) and in the infra-red (H_2O, CO_2, CH_4, CO).

Typical absorption spectra for the molecules occurring in the atmosphere are shown in Fig. 2.3 together with the complete solar spectrum. In the visible there is some absorption due to molecules of oxygen at about 760 nm. In the infra-red, up to about 7 μm, there is strong absorption due to molecules of carbon dioxide and water. Most of this absorption occurs in the lower atmosphere, below about 50 km, where water vapour and carbon dioxide are largely concentrated. Beyond 7 μm there is a fairly transparent region out to about 13 μm, except for an absorption peak near 10 μm due to ozone. This is the so-called atmospheric window and it is the region in which infra-red devices, e.g., heat seeking missiles, must operate if they are to be effective. Beyond 14 μm, solar radiation is almost totally absorbed by the rotational bands of H_2O and by the 15 μm CO_2 bands.

Our treatment of solar radiation and its passage through the atmosphere has of necessity been a very simple one. In particular, we have neglected the fact that the amount of radiation received from the sun will depend on the time of year and the location on the globe at which the measurement is made. Climate will have a considerable effect on the amount of radiation actually reaching the earth's surface, for example, through the amount of cloud cover.

In the first instance, the solar radiation reaching the top of the atmosphere undergoes a slight variation due to the particular motion of the earth about the sun. The earth's orbit is an ellipse with the sun at one of the focal

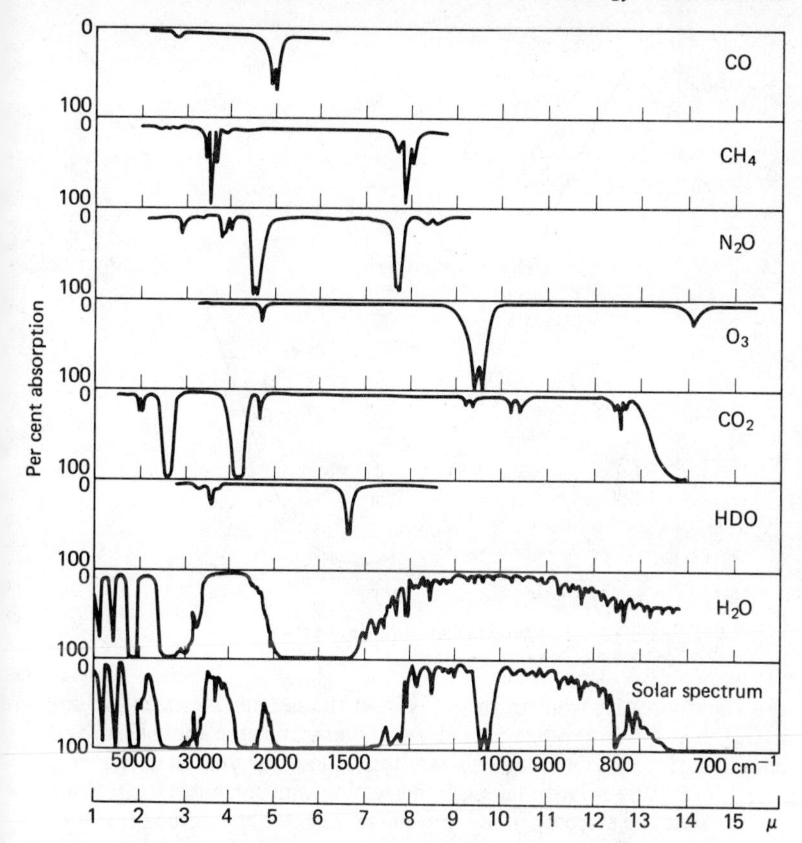

Fig. 2.3 Typical absorption spectra for the molecules occurring in the atmosphere and the complete solar spectrum. (J. T. Houghton and S. D. Smith, *Infra-red Physics*, 1966. Reproduced by permission of the Clarendon Press, Oxford.)

points. This means that the earth–sun distance varies throughout the year, reaching aphelion, a maximum of 152.1 million kilometres in early July, and perihelion, a minimum of 147.1 million kilometres in early January. Fig. 2.4 illustrates the resulting variation in the solar constant during the year. The overall variation in radiation received is seen to be some 6 or 7%.

This effect is swamped, however, by the changes in the pattern of radiation received over the globe during the year as a result of the earth's axis being inclined at other than 90 ° to the plane of its orbit about the sun. Obviously the earth's globe shape means that when the sun is overhead at the equator, the radiation received per unit surface area, the 'insolation on a horizontal plane', at latitude θ° north or south is $\cos\theta$ times the insolation at the equator, if we neglect atmospheric effects (Fig. 2.5). But as the earth is inclined at an angle of $66\frac{1}{2}^\circ$ to the plane of its orbit (Fig. 2.5) the 'effec-

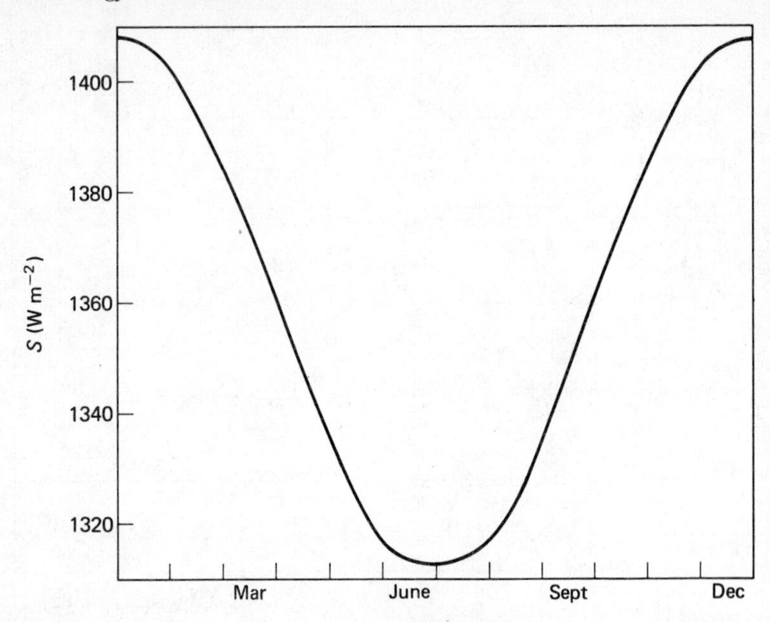

Fig. 2.4 The variation in solar constant throughout the year.

tive equator' varies from latitude $23\frac{1}{2}$°N at the summer solstice (21 June) to $23\frac{1}{2}$°S at the winter solstice (21 December). Thus the energy received on unit surface area per day will vary throughout the year as shown in Fig. 2.6 for various latitudes, again neglecting atmospheric effects. If we now include the effects of the atmosphere, as described in the previous section, we of course reduce all the values of radiation received at the ground. Fig. 2.7 shows the average annual distribution with latitude of solar

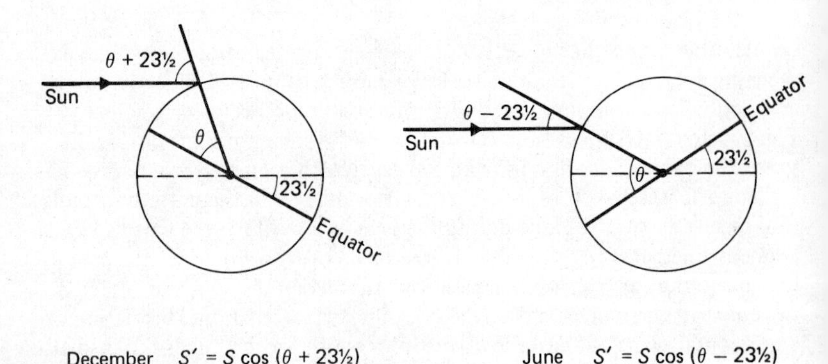

Fig. 2.5 The variation of effective solar constant with latitude.

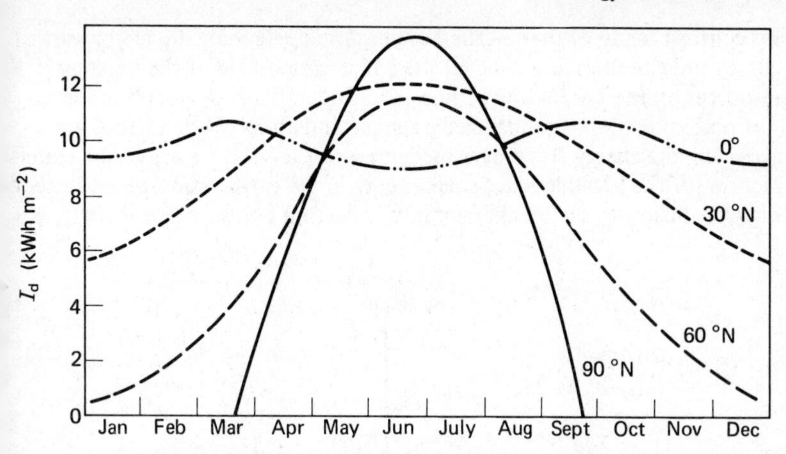

Fig. 2.6 Latitudinal variation of daily insolation (neglecting atmospheric effects).

radiation at ground level (the secondary maximum at the South Pole is due to the extremely clear Antarctic atmosphere).

These seasonal and geographical variations have great importance for climate and for agriculture. It is also worth emphasising the extremely obvious point that they directly influence the amount of fuel which we must use in order to keep warm. During the winter months, European and North American countries have to expend large quantities of their fuel resources simply in warming buildings. (For example, in the United Kingdom some 40% of power production is used for domestic heating in the winter months.)

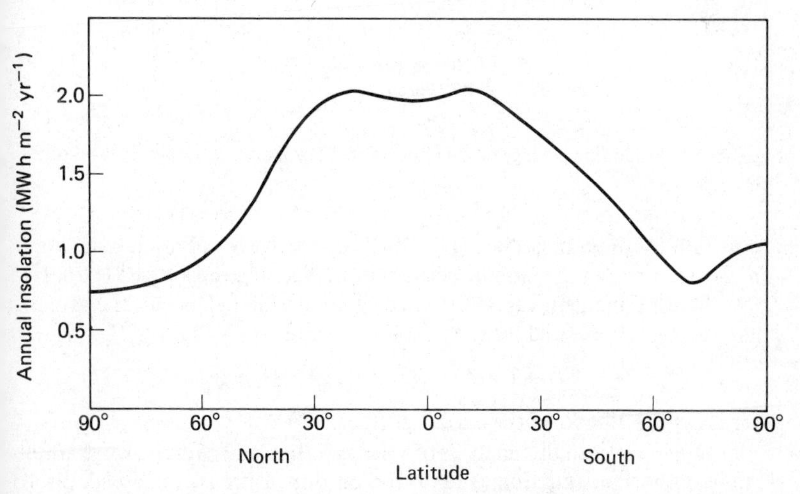

Fig. 2.7 The average annual insolation received at the ground as a function of latitude

By contrast many of the so-called under-developed countries are in warmer climes and can therefore concentrate a high proportion of their energy resources on the production of power for industrial and agricultural use.

Combining these seasonal, geographical and atmospheric effects, and averaging the energy flows over the course of the year, we arrive at a simple picture for the breakdown of solar energy in the earth–atmosphere system. Fig. 2.8 illustrates the breakdown into reflection by the atmosphere (Q_r),

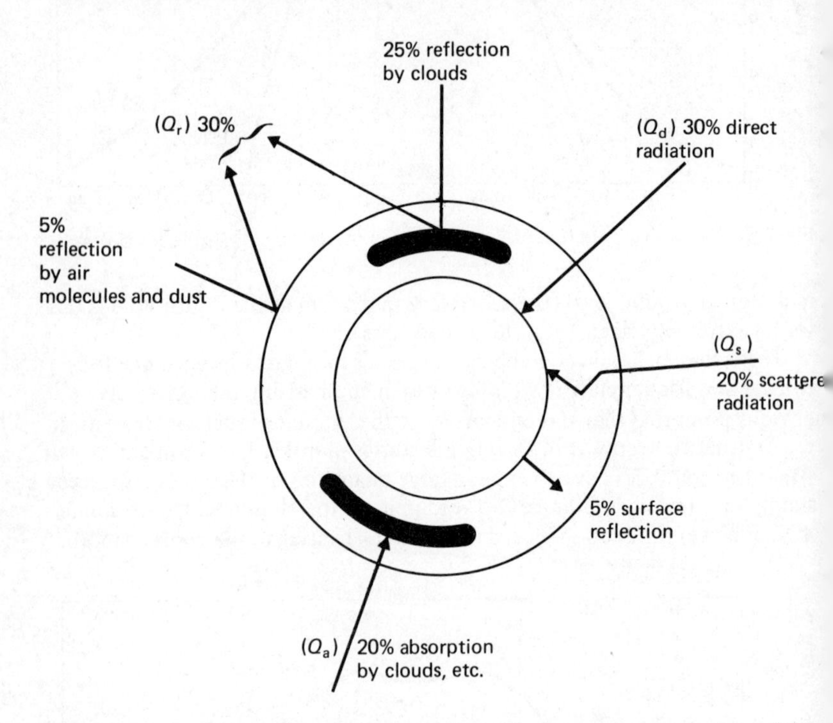

Fig. 2.8 The radiation balance of the atmosphere. The figures show percentages of the total solar input.

absorption by the atmosphere (Q_a), diffuse shortwave radiation reaching the earth's surface (Q_s), and direct radiation reaching the surface (Q_d). If Q_T is the total incident radiation, then the radiation balance in the atmosphere can be represented by the simple equation

$$Q_T = (Q_d + Q_s)(1 - A) + (Q_d + Q_s)A + Q_r + Q_a$$

where A is the albedo of the earth's surface.

Turning now from this analysis of solar radiation and climate, we examine its most important function as far as life on this planet is concerned, photosynthesis. All of the chemical energy resources, the fossil fuels, which man uses could not have existed but for photosynthesis, and indeed life as we know it on this planet owes its survival (and its origin) to the photosynthetic process, the trapping and storage of solar energy by green plants.

The basic energy cycle of life has in effect three phases. These are the trapping and storage of the sun's energy, the extraction of energy from the store and the utilisation of this energy. Plant cells assimilate the sun's energy by the process of photosynthesis, which takes place in the chloroplasts of green leaves. The solar energy is stored as potential chemical bond energy by the reduction of carbon dioxide to form large molecules such as carbohydrates, lipids and proteins. In building these large molecules, many new chemical bonds are created, and it is in these bonds that the converted solar energy is stored. (One way of looking at this is to say that the atoms have been arranged in a more ordered fashion in the large molecules and, according to the laws of thermodynamics, this requires energy.) Plant and animal cells can extract the energy stored in fats, carbohydrates and proteins for their own needs through the process of respiration. It is interesting to compare biological and physical systems here. In the experience of the physicist or engineer the most common way of transferring energy is by some thermal process, but as most biological systems are isothermal they have developed a mechanism of energy transport by exchanging chemical bonds, or electrons, together with their energy, from one molecule to another.

When a cell requires energy it breaks down the large polymeric molecules produced by photosynthesis. Instead of releasing the bond energies as heat, it does so by carefully controlled reaction sequences in which the energy is re-stored in smaller, more manageable molecules. These small molecules can now be sent off to a particular region of the cell where energy is required. The most commonly occurring of these mobile, high-energy molecules is adenosine triphosphate, or ATP for short, whose structure is shown in Fig. 2.9. Some measure of its importance can be gauged from the fact that each DNA molecule (deoxyribonucleic acid, the genetic codebook of any cell) requires some 120 million molecules of ATP *per second* to provide the energy necessary for the synthesis of major cell components. In a neutral solution, that is at a pH of 7, the ATP molecule has a highly negative charge, as depicted in Fig. 2.9. It is in forming this high concentration of negative charge in the pyrophosphate bonds that work must be expended and so, in going from ADP, the diphosphate, to ATP, energy is stored in the unstable terminal link. Breaking down this link enzymatically, in a controlled fashion, releases the stored energy, and the hydrolysis of ATP to ADP and phosphoric acid yields some 30 kilojoules (7 kcal) per mole, equivalent to 0.3 eV per molecule.

Returning to photosynthesis itself, we should remember that not only does it convert the sun's energy into a manageable form for the plant, but it also maintains the oxygen balance in the atmosphere. The overall reaction occurring in photosynthesis is the fixation of carbon dioxide and water to

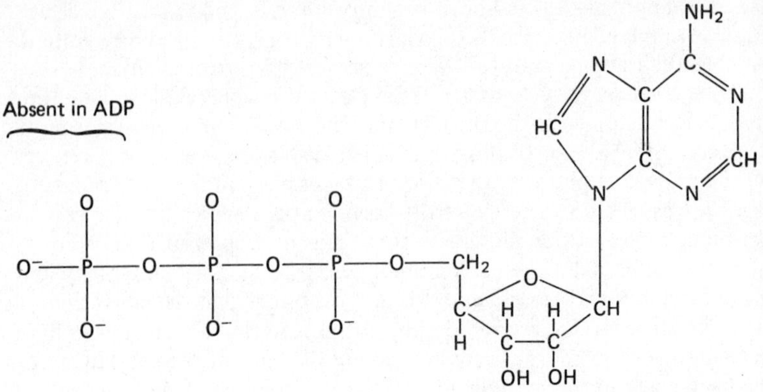

Fig. 2.9 The molecular structure of adenosine triphosphate, ATP.

form a sugar and oxygen. We can grossly oversimplify the process and represent it by

$$6CO_2 + 6H_2O + nh\nu \rightarrow \underset{\text{(sugar)}}{C_6H_{12}O_6} + 6O_2$$

where $h\nu$ is a quantum of light energy absorbed by the plant, and n gives us a measure of the overall efficiency. As we have mentioned, the reaction centres are the so-called chloroplasts contained in green vegetation. Fig. 2.10 shows a typical plant chloroplast, of which there might be 50 in each cell, and emphasises the lamellar arrangements (grana) of the absorbing pigments within them. Chloroplasts can, of course, be extracted and biologists have even managed to cause them to photosynthesise in an artificial environment! It is the lamellar arrangements, the grana, which actually contain the pigments responsible for absorbing the sun's light – chlorophyll, carotenoid, phycobilin.

Of these pigments only chlorophyll exists in all photosynthetic cells. It is found in two forms: chlorophyll a, in all green plants, and chlorophyll b, in most green plants. These molecules have somewhat complicated structures (Fig. 2.11) in which there is a porphyrin structure, similar to that found in haemoglobin, but with a magnesium instead of an iron atom at its centre. Attached to this hydrophylic porphyrin ring is a long hydrocarbon side chain which is hydrophobic – so chlorophyll should be a detergent! By looking at the way in which chlorophyll absorbs light (Fig. 2.12), we can immediately understand the green colouration of leaves: strong absorption in the blue and red regions of the visible spectrum means that a predominantly green colour will be transmitted.

The chemistry of photosynthesis is extremely complicated. The conversion of carbon dioxide molecules into sugar requires some 5 eV per molecule, which is significantly greater than the energy carried by each quantum of visible light (just under 2 eV for red light, approximately 3 eV for blue

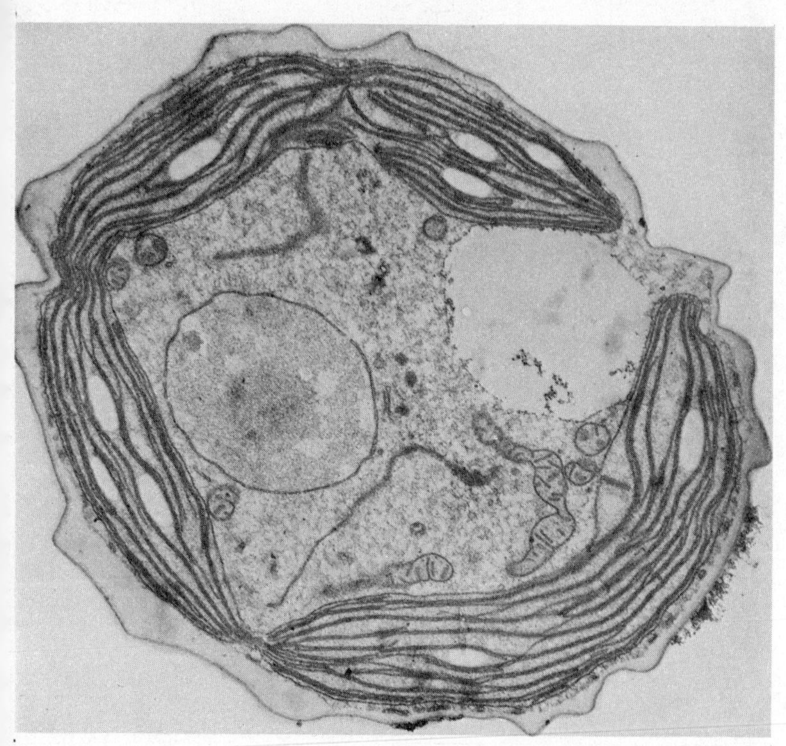

Fig. 2.10 Transmission electron micrograph of section through a cell of unicellular green algae *Chlorella pyrenoidosa.* (Kindly provided by Dr R. Marchant, New University of Ulster.)

light). So although we know that visible light is essential, the process cannot be a direct one involving individual quanta. Similarly, although we know that the same number of oxygen molecules is produced as carbon dioxide molecules are consumed, it is found by using radioactive O^{18} tracer atoms that the oxygen does not come directly from the carbon dioxide, but from the splitting of water molecules. On top of this, the energy needed for fixing the carbon dioxide and splitting the water molecules is supplied by ATP molecules, themselves generated from ADP and inorganic phosphate (the process called phosphorylation) within the chloroplasts. The overall action of the chloroplasts is illustrated (grossly oversimplified) in Fig. 2.13. It is worth noting that there is no instantaneous balance between fixation of carbon dioxide and oxygen released, but over a period of time this balance is ensured by a built-in feedback mechanism. For if carbon dioxide fixation proceeds more rapidly than oxygen evolution, more ADP is available to speed up the rate of phosphorylation, and vice-versa.

Of course, in the context of a book on energy resources, our main inquiry about photosynthesis must concern its efficiency. Fig. 2.14 shows the

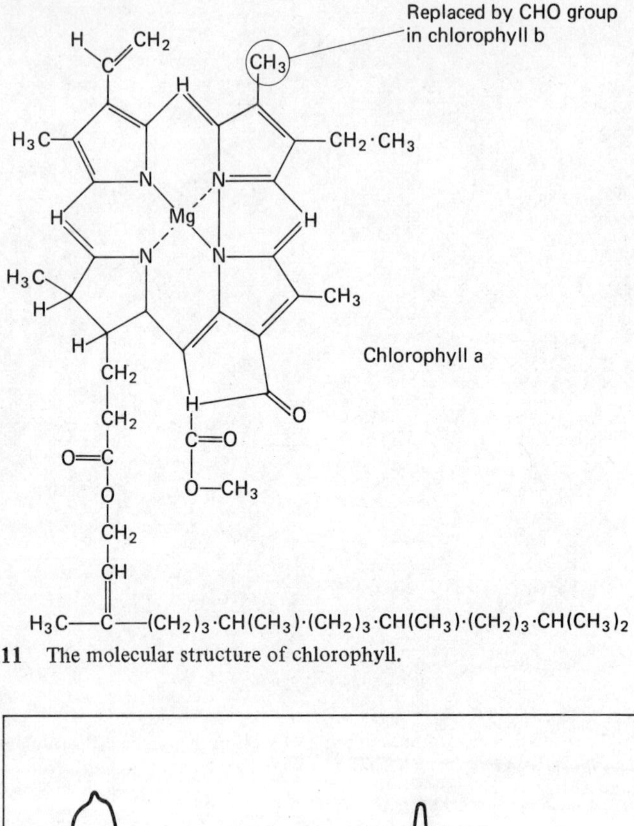

Replaced by CHO group in chlorophyll b

Chlorophyll a

$H_3C—C—(CH_2)_3 \cdot CH(CH_3) \cdot (CH_2)_3 \cdot CH(CH_3) \cdot (CH_2)_3 \cdot CH(CH_3)_2$

Fig. 2.11 The molecular structure of chlorophyll.

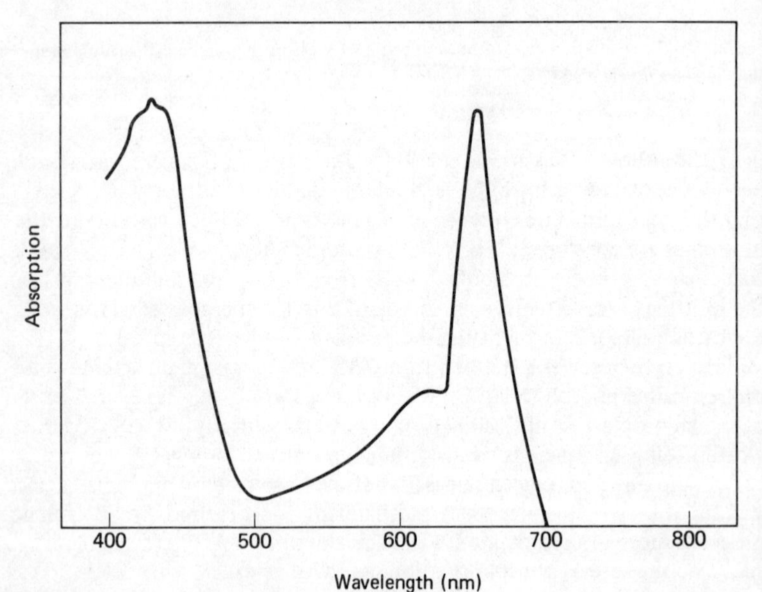

Fig. 2.12 The absorption spectrum of chlorophyll.

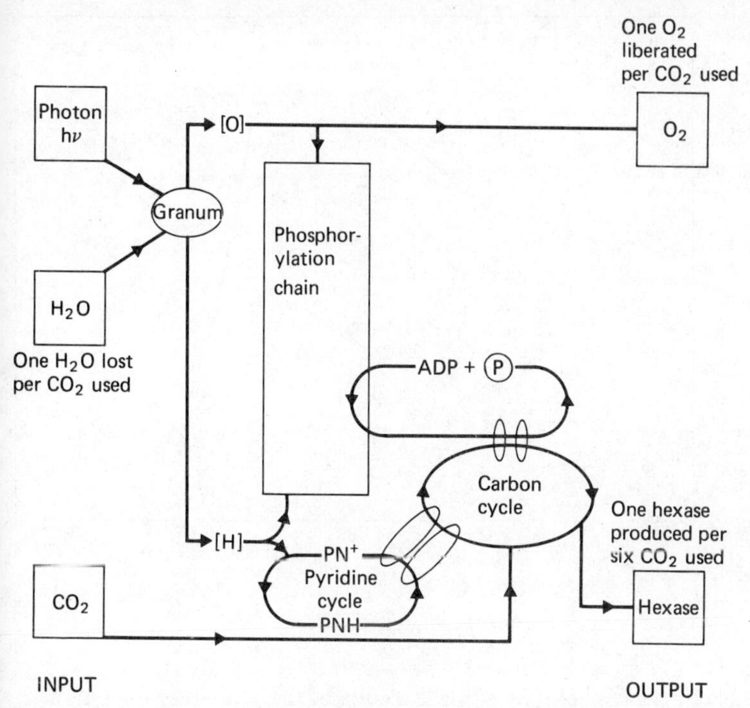

Fig. 2.13 A schematic representation of the overall action of chloroplasts.

'action spectrum' (yield of sugar per quantum of light as a function of wavelength) for a single-celled green algae called Chlorella. This curve is extremely flat in comparision with the absorption spectrum of chlorophyll (Fig. 2.12), implying that light quanta other than those used by chlorophyll are effective in photosynthesis. This suggests the existence of other photochemical pigments and, in addition, from studies of the rates at which the various chemical reactions proceed in photosynthesis, it would seem that stable intermediate compounds are formed in the reaction sequence.

The first measurements of quantum efficiency in photosynthesis were made by Warburg and Negelein in 1922 when they showed that four quanta of light of any wavelength in the visible range were necessary for the reduction of one molecule of carbon dioxide to carbohydrate. Since then the measurement of this efficiency has caused much controversy because of the problems of correcting for respiration at the very low light intensities used in such experiments. The accepted result now is that at least eight quanta of light energy are required to reduce one molecule of carbon dioxide. However, the overall efficiency of green plants in utilising solar energy is not nearly so good as we might imagine from these figures.

We have seen that about 25% of solar radiation reaching the ground is

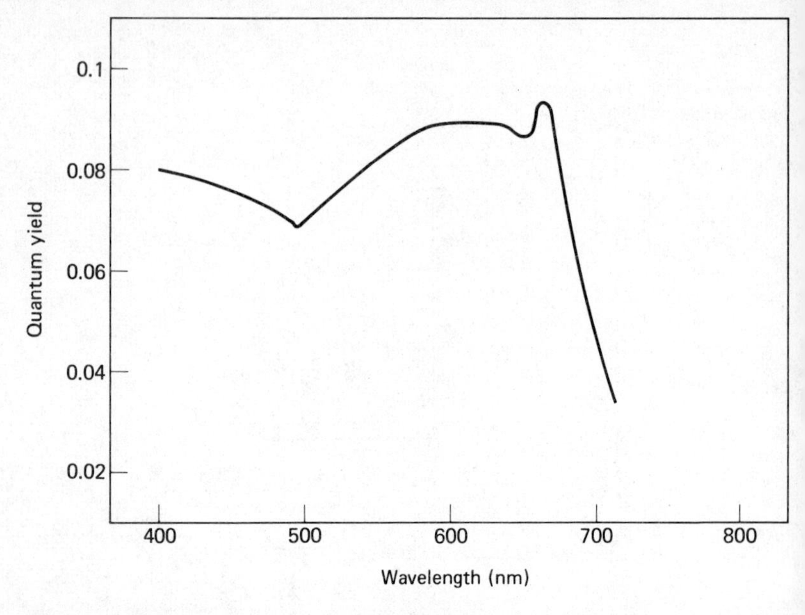

Fig. 2.14 The action spectrum of chlorophyll.

useable by plants, but only a small fraction of this is actually used in the photosynthetic process. Thus man's most efficient crop as far as photosynthesis is concerned, sugar cane, only uses an average of 1.4% of the available solar radiation during the year. A forest, by comparison, may only achieve an efficiency of 0.5%. Many factors such as temperature, carbon dioxide concentration, or the availability of water and nutrients in the soil, can limit this 'primary productivity'. Indeed, plants can be made to increase their primary productivity by raising the concentration of carbon dioxide, and will go on doing so until at least three times the normal concentration of the gas has been reached.

Various attempts have been made to estimate the primary productivity of plants in different regions of the world. This can be done by measuring the dry weight of annual crops when they are harvested at the end of the growing season and taking this as a measure of the total carbon fixed. At first sight the most surprising result would seem to be the small variation with latitude in the maximum rate of photosynthesis during the growing season. For example, a Minnesota maize crop, in summer, will fix carbon at a rate of about 8 grams per square metre per day, compared with a figure of about 7.5 for a tropical rain forest. Averaged over the whole year, however, the rain forest is twice as productive as the maize crop. Similarly, the maximum rate of photosynthesis in Arctic regions is comparable to that in much warmer climes. This is simply a result of the greater amount of sunlight available to Arctic plants during the growing season because of the longer day.

Can we improve the efficiency of photosynthesis? Improvements in the productivity of crops would be most welcome, especially in densely populated countries such as the United Kingdom, which is not, at present, self-sufficient in food. We mentioned above that a typical crop such as maize will use only about 1% of the radiation it receives from the sun during the year and, when we realise that only about 25% of the total dry mass of the crop can be consumed, we can see that the overall efficiency of food production is a meagre 0.25%. A ten-fold improvement in this efficiency is well within the limits imposed by the photosynthetic process.

The principal reason for the low yields from crops is that they are annuals, which means that for some time after sowing there are no leaves to absorb sunlight and peak photosynthetic efficiency is only achieved for a relatively short period in the growing season. We could improve this by setting up our agriculture on a continuous basis so that there would be a leaf canopy throughout the year. Harvesting would then be a continuous process involving the collection of the older leaves, which could be treated to yield protein.

We have already mentioned that yields can be improved by increasing the ambient concentration of carbon dioxide, but this is hardly going to be applicable on an agricultural scale.

Another possibility is the mass culture of unicellular algae, such as chlorella, which have a high photosynthetic efficiency under laboratory conditions. As an additional advantage, 60% of their dry weight is high-quality protein. A pilot plant is, in fact, already in operation at the Micro-algae Research Institute of Japan, near Tokyo. The algae are cultured in large ponds with carbon dioxide continuously pumped through the suspension. Typical daily yields of about 4 grams carbon per square metre have been obtained which, although good, are still a factor of two or more down on the yields of maize or sugar cane. Nevertheless, the *protein* production of these chlorella ponds is much greater than that for conventional crops. The main problem with this technique at the moment is its very high cost.

Perhaps the most hopeful line of attack is to use our extensive knowledge of genetics and cross-breeding of strains to produce a crop with a high photosynthetic efficiency. In this respect a recent discovery may have great significance. This concerns the herbaceous perennial, Tidestromia oblongifolia, which is to be found in large numbers on the floor of Death Valley, California, one of the harshest environments on earth. Whereas most plants in Death Valley grow only in the milder winter months, most of Tidestromia's photosynthetic activity takes place from May to August, the hottest and driest months of the year! To survive these conditions, the plant has developed a highly efficient photosynthetic apparatus which can still function at high temperatures and low moisture levels. If suitable crops could be bred incorporating this characteristic, the productivity of poor agricultural land could be greatly enhanced.

Apart from the creation of conditions conducive to the development of man himself, the most important legacy from the photosynthesis of bygone ages is, of course, the fossil fuel bank on which we are now drawing so heavily. How were these fossil fuels formed and what estimates can we make as to the likely magnitudes of the resources?

All the fossil fuels – coal, peat, oil and natural gas – are alike in that they are the remains of decayed organic matter, but the nature of the organisms, the conditions under which the decay occurred and those under which the fuel accumulated are different for the different fuels.

It is fairly well established that coal and peat arise from vegetation which grew on wet land such as swamps in river estuaries. When the vegetation died, it fell into the swamp and decayed under almost air-free conditions by microbiological processes which can be summarised very approximately by the equation

$$\underset{\text{cellulose}}{2C_6H_{10}O_5} \rightarrow \underset{\substack{\text{humified}\\ \text{residue}}}{C_8H_{10}O_5} + 2CO_2 + \underset{\text{methane}}{2CH_4} + H_2O$$

The product of this reaction, together with compounds such as lignin which resist decomposition, forms peat.

The transformation of peat into coal involves burial by sandy sediments, and it is thought that this might have occurred most easily in a river delta if some sort of subsidence of the land took place, allowing the sea to sweep over the swamp. It appears that such a process often occurred many times, producing a series of coal seams at different depths.

Geological evidence indicates that most of the coal which is mined in Britain, and much of that mined elsewhere, is the remains of the lush vegetation of the Carboniferous period around 300 million years ago. It is fairly certain that the climate at that time was considerably more tropical than it is today in the coal-mining regions, an effect which has been attributed to various causes, including the relatively recent theory that the continents are drifting around the surface of the earth. Whatever the reason, the result is that the rate at which coal-forming material accumulated at that time was considerably greater than it is today in the same places. There is some doubt as to whether or not the overall rate of accumulation of coal in the world today is less than it was then, but even the most optimistic view would not suggest that it is any higher. What is certain is that the rate of accumulation is so much less than the rate of consumption that for all practical purposes coal must be regarded as an exhaustible source of energy. It should be emphasised at this point that despite the finite nature of the resource, there is nothing necessarily wrong in using it. It may be that our heirs some centuries from now will find themselves without coal as a fuel because we have burnt it; however, if we act responsibly now in our approach to energy supplies, they may well not need it as a fuel.

What are our reserves of coal, that is how much *mineable* coal is there in the world? An exact answer is impossible, but a recent estimate by the United States Geological Survey is about 7.6×10^{12} tonnes. In 1928 the same organisation quoted a figure of about 3.2×10^{12} tonnes. Part of the difference between these figures arises from a change in the definition of what is mineable. The earlier figure excluded coal at depths greather than 3000 feet (915 metres), while the later estimates including deposits at 4000 or even 6000 feet (1220 or 1830 metres). This difference points out two most important factors in estimation of reserves, both of which tend to be forgotten: any estimate is to some extent arbitrary, since it depends on the definition of what is useable, and any estimate is somewhat inaccurate, since it depends on the state of knowledge at the time. As a rough rule of thumb, it is fair to assume that if two estimates differ by no more than a factor of two, they agree accurately, and must not be regarded as evidence of inconsistency. Accepting this proviso, and taking the larger of the two estimates from the US Geological Survey as our basis for calculation, the coal resources of the world represent a store of energy of about 2×10^{23} joules or 10^{20} watt hours.

The regional distribution, shown in Table 2.1, is interesting because all the major industrial nations of the world appear to have adequate coal reserves within their own territories. To some extent, of course, it can be argued that this was a prerequisite for their industrialisation. While coal is

Table 2.1 World coal resources in tonnes $\times 10^9$

Country	Hard coal	Lignite
Australia	16	95
Brazil	11	–
Canada	61	24
China	1011	–
Colombia	13	–
Czechoslovakia	12	10
Germany (East)	–	30
Germany (West)	70	–
India	106	–
Japan	19	–
Poland	46	15
South Africa	72	–
USSR	4121	1406
United Kingdom	16	–
USA	1100	406
Yugoslavia	–	27

From United Nations Statistical Year Book (1972). Copyright of the United Nations, 1972. Reproduced by permission.

not as essential today, it is still true that the major coal-consuming nations are for the most part the major producers. Consequently, despite the vast importance of coal as a fuel, it is not a similarly vast item in international trade.

Peat is of much lower importance as a fuel than coal, mainly because it is less satisfactory to handle on a large scale, and also because of its low calorific value – about 11 500 kJ kg^{-1} (5000 Btu lb^{-1}) compared to about 23–32 500 kJ kg^{-1} (10–14 000 Btu lb^{-1}) for coal. The main problem with peat is the high water content, which even after drying is as high as 40–50% by weight. This leads to a low calorific value and causes handling difficulties Nevertheless, in countries such as the USSR, where there are very large amounts, and Ireland, where there are few other resources of fuel, peat is used on quite a substantial scale. The regional distribution is shown in Table 2.2 where the figures are given in percentages rather than absolute terms because of the difficulties of accurate estimation. Two recent studies of world peat resources have given figures of 3.3×10^{11} and 2×10^{11} tonnes (dry weight), which, bearing in mind the comments made earlier about coal, can be seen as agreeing well with each other. Again taking the larger figure, this represents a store of energy of about 7×10^{22} joules. Comparing the figure with that for coal, it can be seen that it is less but not very much so – in other words peat deserves to be regarded as a major fuel resource. Unfortunately it is uncertain how much of the world's peat can usefully be used as fuel; much of it is upland 'blanket' peat which is difficult to harvest on a large scale.

The rate of accumulation of peat has been estimated at the remarkably large figure of three tonnes per hectare per year. On this basis it would

Table 2.2 World peat resources

Country	Per cent of world's resources
USSR	60.8
Finland	9.5
Canada	9.1
USA (excluding Alaska)	5.0
Germany (GDR and FRG)	3.8
Great Britain and Ireland	3.5
Sweden	3.4
Poland	2.3
Indonesia	0.9
Norway	0.7
Cuba	0.3
Japan	0.2

From A. S. Olenin, Transactions of the Second International Peat Symposium, *Leningrad 1963. Reproduced by permission of the Controller, HMSO.*

appear that it is not only a large source of fuel, but a renewable one as well. Unfortunately it does not usually accumulate conveniently, but rather is distributed over a large land area. It is probably true to say that at the present rate of worldwide consumption (9×10^7 tonnes per year), it is being consumed faster than it accumulates in a useable manner. Accordingly, it is fair to regard peat as a non-renewable resource, but it is much less so than the other fossil fuels.

We turn now from the solid fuels to the petroleum fuels, oil and natural gas. The origin of petroleum is rather different from that of coal and peat. It arises from aquatic organic matter, mostly green plankton, which accumulated in the sediments at the bottom of the sea and decomposed anaerobically, leading to a dispersion of droplets of oil and water between the grains of sedimentary rock. Uusually the rock in which the oil is formed, termed the source rock, is of a fine-grained structure from which extraction is difficult, and, before the dispersed oil can come together to form an exploitable reserve, it must travel to a coarser-grained rock, termed the reservoir rock, in which it can accumulate and from which it can be extracted at a reasonable rate. The mechanism by which the oil moves can be attributed to entrainment of the dispersed oil droplets in circulating water. If nothing intervenes, the oil will continue to move until it reaches the surface as a seep. A more useful situation arises when a structure such as that sketched in Fig. 2.15 intervenes. Here a coarse-grained reservoir rock is overlain by a fine-grained impervious rock known as a cap rock. The whole structure is termed an oil trap. Once within the trap the oil, being less dense, separates from the water and floats on top; there may also be separation of the gaseous petroleum fractions as well.

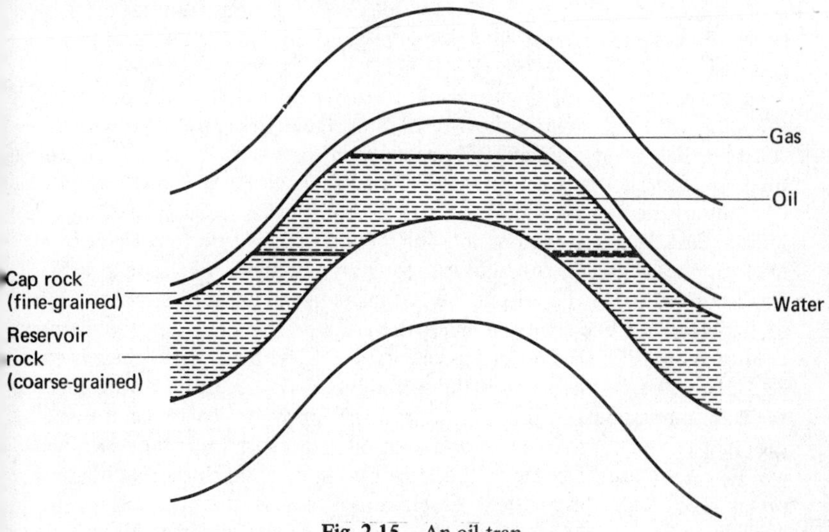

Fig. 2.15 An oil trap.

As in the case of coal, it is probable that the processes by which oil resources form are still continuing, but at a rate very much less than the rate of exploitation, so that oil must be regarded as exhaustible. The same applies to natural gas. There is a complication in the estimation of world oil supplies which also applies to some extent to coal, but less forcibly. It is a very expensive operation to locate new oil deposits, much more so than for coal because useful deposits occur at very great depths. Consequently the oil industry does not take the trouble to survey possible wells accurately until there is a reasonable prospect of sales from the new well within the the next fifteen years or so. Reserves whose potential has been accurately assessed are called *proved reserves.* Clearly the proved reserves are always less than the total amount of oil known, and in view of the uncertainties involved in locating and surveying deep wells, the total amount of oil known is likely to be considerably less than the actual amount which may eventually be found. It is all too easy, therefore, to portray the diminishing supplies of oil in the world in terms of the imminent collapse of civilisation. In reality that is unlikely. Nevertheless, there is some cause for concern, for it has been estimated that world oil reserves amount to no more than 2×10^{12} barrels – about 10^{22} joules. Gas reserves associated with the oil contribute about the same amount of energy, roughly doubling the total to 2×10^{22} joules. There is an additional resource in oil shales and tar sands, which may well amount to more than the total of liquid oil, but how much of this can be economically exploited is uncertain. Thus it is evident that oil reserves are appreciably less than those of coal, about 10% in round figures. In view of the essential nature of oil for all sorts of purposes, both fuel and petrochemical, it is of some importance to look for possible ways of shifting some of the burden of world energy demands from oil on to other sources such as coal, nuclear power, and the so-called natural power sources.

A further reason for the uncertain future of oil supplies, and one which is actually of much greater short-term importance than any worldwide shortage, lies in the geographical distribution, shown in Table 2.3. The most obvious difference between this list and that for coal is the complete absence of many of the world's industrial nations, and the vast predominance of the Middle East. It is quite impossible for the industrial world to manage on its own supplies of petroleum, and it is equally impossible for the relatively small population of the Middle East to make use of all of its supplies. Consequently oil, unlike coal, is a major item of world trade. When the member countries of OPEC, the oil producers' cartel, raised their prices sharply in 1973, the oil consumers had little option but to pay, and the world suffered the economic effects of this situation for several years. This aspect of the problem arose not from any shortage of oil – in fact a short-term surplus was brought about because of the reduction in demand caused by higher prices – but from the political association of most of the producers, and, consequently, it could be solved by discovering new sources of oil in

Table 2.3 World oil and gas reserves (proved), December 1976. Oil in barrels $\times 10^9$, gas in cu ft $\times 10^{12}$ Excludes amounts less than 1×10^9 barrels and 10×10^{12} cu ft

	Oil	Gas
Asia		
Australia	1.4	32
Brunei	1.6	–
China	20.0	25
India	3.0	–
Indonesia	10.5	24
Malaysia	2.4	15
Europe		
Netherlands	–	62
Norway	5.7	18
UK	16.8	30
USSR	78.1	918
Middle East		
Abu Dhabi	29.0	20
Dubai	1.5	–
Iran	63.0	330
Iraq	34.0	27
Kuwait	67.4	31
Neutral zone	6.3	–
Oman	5.8	–
Qatar	5.7	27
Saudi Arabia	110.0	63
Syria	2.2	–
Africa		
Algeria	6.8	126
Angola	1.2	–
Egypt	1.9	–
Gabon	2.1	–
Libya	25.5	26
Nigeria	19.5	44
Tunisia	2.7	–
Western Hemisphere		
Argentina	2.3	–
Canada	6.2	56
Equador	1.7	12
Mexico	7.0	12
Venezuela	15.3	41
USA	31.3	220

Source: Oil and Gas Journal, *December 27, 1976. Reproduced by permission of the Petroleum Publishing Co.*

countries with different political policies. It can also be solved, less quickly but in a more lasting manner, by diversifying the sources from which the world gets its energy.

Fossil fuels are so dominant in the world energy scene at present that we are in danger of forgetting that they represent but a tiny fraction of the incoming solar radiation. The same is also true of the photosynthetic process by which the fossil fuels were originally formed. Compared with this, the other manifestations of solar radiation are very large indeed.

The main impact of solar radiation at the earth's surface is in the evaporation of water from seas, lakes and rivers. Nearly one quarter of the total incoming solar power is dissipated in this way. This water rises into the atmosphere, gaining potential energy in so doing, but cooling and forming droplets of water. During precipitation these droplets fall back to the earth's surface, and those which fall on high ground are still left with some useful potential energy. This is partially converted into kinetic energy (some is lost to friction) as the water flows back towards the sea and it is this kinetic energy which can be used to drive mechanical devices such as the turbines in hydroelectric power schemes (Chapter 4).

If a stream has a flow rate of F $m^3 s^{-1}$, and drops through a vertical height H metres (the 'head' of water), then the power available is

$$P = \rho F g H$$

where ρ is the density of water (1000 kg m^{-3}) and g is the acceleration due to gravity (9.8 m s^{-1}). Therefore,

$$P = 9.8\ FH \quad \text{kW}.$$

By studying stream flow records throughout a given country an estimate can be made of its potential hydropower capacity. These estimates are listed in Table 2.4 together with the developed capacities in the regions concerned. Notice that the total estimated world potential at close to three million megawatts is only a fraction of one per cent of the solar power used in evaporation!

Table 2.4 Regional Distribution of Hydropower Resources

	Potential (1000 MW)	Per cent of world total	Developed (1000 MW)
North America	313	11	59
South America	577	20	5
West Europe	158	6	47
Africa	780	27	2
Middle East	21	1	–
South East Asia	455	16	2
Far East	42	1	19
Australia	45	2	2
USSR, China, etc.	466	16	16
	2857	100	152

(Reference: M. King Hubbert, 1969, in Resources and Man, *Publication 1703, Committee on Resources, National Academy of Sciences – National Research Council, W. H. Freeman and Co., San Francisco, 1969. Reproduced by permission of the National Academy of Sciences.)*

By comparison with water power, wind power is a mere drop in the ocean. Fig. 2.1 suggests that only 370 million megawatts of power are available from the winds and waves across the globe, and much of this energy is dissipated in inaccessible regions. Nevertheless, the available power is still more than enough to satisfy most of man's requirements if only it could be tapped.

In theory, the power available to a windmill of unit cross-sectional area sitting in an air flow of velocity v (m s^{-1}) is 0.65 v^3 watts. However, the efficiency of such a windmill can at best be approximately 60%. (This is an aerodynamic limitation arising from the deflection of the air flow by the windmill.) This leaves us with a theoretical power extraction of $0.4v^3$ W m^{-2} Because of this cubic relationship, data for average wind speed cannot be used to predict average wind power (this is because the average of v^3 is not equal to the cube of the average of v). However, it can be shown that the substitution of v in the expression for power always leads to an underestimate of the average power available. In a windy region such as the British Isles mean wind speeds are of the order of 10 mph (4.5 m s^{-1}) or greater and are often considerably in excess of this in coastal areas. This mean value leads to a theoretical power availability of at least 35 W m^{-2}.

The countries of Western Europe are also well placed to take advantage of wave power. In the North Atlantic ocean the prevailing winds are from the west and cause a considerable swell in the coastal waters around the United Kingdom and in the North Sea. Even if there is no wind blowing locally, ocean waves are probably being generated somewhere in the ocean and thanks to the long attenuation distance (of the order of 5000 metres), and to the long decay time (more than twelve hours), a virtually continuous supply of wave energy is available. On average each metre of wavefront could provide 80 kW of power, so that the total coastline of the United Kingdom could yield more than 100 000 MW. This is significantly greater than present electricity generation! However, there are mammoth technical problems and realistic efficiencies have yet to be established. The problems associated with wind and wave power will be discussed in Chapter 4.

To complete the picture for solar derived sources of energy we really ought to consider power extraction from ocean currents and temperature gradients. However, at the present time these estimates are even more in the realm of guesswork than for other energy sources. Moreover, howls of protest have already been heard in the British Isles and Northern Europe over a proposal to extract energy from the Gulf Stream by placing turbines in its path off the coast of Florida. Perhaps these estimates, as well as the energy source, are best left alone.

Referring once again to Fig. 2.1 we see that the power available from tidal sources is estimated to have a potential of 3 million megawatts. The tides, that is the rise and fall of the water level around our coasts twice each day, arise primarily from the gravitational attraction between the earth and the moon. In mid-ocean the tidal range, the difference between the highest and lowest levels, is only a metre or less, but in some coastal estuaries it can be much greater. This is due to the amplification of the tidal wave as it sweeps up the narrowing channel, its energy concentrating into a decreasing volume of water. It can also be the case in an enclosed sea, such as the Bay of Fundy, that resonance conditions occur in which case a large tidal range will be observed.

An upper limit to the available tidal power is the energy dissipated in the shallow seas and estuaries around the earth's coastline. This is estimated to be one-third of the total tidal power dissipation, that is, 10^6 MW, but the true potential is probably much less than this and to estimate it we must consider briefly the way in which tidal power is extracted.

Basically a tidal power installation involves impounding water in an artificial basin at high tide and then allowing it to escape at low tide. In a similar fashion to hydroelectric schemes, the escaping water can drive turbines and generate electrical power. The simplest way of making such an artificial basin is to dam the narrow mouth of an estuary. If A is the surface area of the basin and R is the tidal range, then the potential power extraction per tidal cycle (approximately twelve hours) is $\frac{1}{2} \rho g A R^2$, where ρ is the density of water and g is the acceleration due to gravity. Using this formula the potential power and energy outputs for a given estuary can be estimated. The results for the most promising sites around the world are listed in Table 2.5. For technical reasons (see Chapter 4) it is likely that the efficiency of a tidal power scheme will be no better than 25%, in which case the data in Table 2.5 must be reduced considerably. This means that the true potential of tidal power may be only of the order of 15 000 MW, approximately 1% of the total available hydropower.

Fig. 2.1 also indicates that the total flux of geothermal heat to the earth's surface is about 32 million megawatts. The greater part of this flux occurs by conduction but the low temperatures involved at any reasonably accessible depth preclude large scale energy conversion. Occasionally hot springs or geysers (and molten lava) manage to reach the surface and the thermal energy of these can be used directly for heating. Alternatively the steam can be used to generate electricity (Chapter 4).

It is possible to place an upper limit on the extraction of power from geothermal sources. Conduction of heat to the surface proceeds at an average rate of 0.06 W m^{-2} (which corresponds to the above figure for total flux) but hot springs and geysers account for only 1% of the total flow. It has been estimated by White (1965) that the useable stored energy in the major fields amounts to 4×10^{20} joules. If we allow a 25% conversion efficiency to electricity this stored energy is equivalent to an output of 60 000 MW over a period of 50 years. The power rating during this period is equivalent to that for tidal power but is only a few per cent of the potential hydropower capacity. If the whole geothermal reservoir can be tapped (a possible technique is described in Chapter 4) then the resource has a capacity of approximately 10^6 MW for a prolonged period of time.

To complete our discussion of world energy resources we now turn our attention to the supplies of nuclear fuels. Estimates of resources are subject to the usual hazards and inaccuracies which have been mentioned in connection with the fossil fuels. There is an added problem that the amount of useful energy that can be extracted from them depends very sensitively on

Table 2.5 Regional Distribution of Tidal Power Resources

	Mean range (m)	Basin area (km^2)	Mean power (MW)	Potential Annual production (1000 MWh)
N. America				
Passamaquoddy	5.5	262	1 800	15 800
Cobscook	5.5	106	722	6 330
Annapolis	6.4	83	765	6 710
Minas-Cobequid	10.7	777	19 900	175 000
Amherst Point	10.7	10	256	2 250
Shepody	9.8	117	520	22 100
Cumberland	10.1	73	1 680	14 700
Petitcodiac	10.7	31	794	6 960
Memramcook	10.7	23	590	5 170
S. America				
San Jose	5.9	750	5 870	51 500
England				
Severn	9.8	70	1 680	14 700
France				
Aber-Benoît	5.2	2.9	18	158
Aber-Wrac'h	5.0	1.1	6	53
Arguenon	8.4	28	446	3 910
Frenaye	7.4	12	148	1 300
La Rance	8.4	22	349	3 060
Rotheneuf	8.0	1.1	16	140
Mont St Michel	8.4	610	9 700	85 100
Somme	6.5	49	466	4 090
Ireland				
Strangford Lough	3.6	125	350	3 070
USSR				
Kislaya	2.4	2	2	22
Lumbouskii Bay	4.2	70	277	2 430
White Sea	5.65	2 000	14 400	126 000
Mezen Estuary	6.6	140	1 370	12 000
			62 125	562 553

(Reference: M. King Hubbert in Resources and Man, *Publication 1703, Committee on Resources, National Academy of Sciences – National Research Council, W. H. Freeman and Co., San Francisco, 1969. Reproduced by permission of the National Academy of Sciences, Strangford Lough data calculated by the authors.)*

the technology used to extract it. This in turn means that the value in terms of net energy of a low-grade deposit can change from strongly negative (using more energy to extract it than is returned) to strongly positive simply by a change in nuclear engineering.

A recent estimate of world uranium reserves by OECD puts the amount available at a price of less than 10 US dollars per pound weight of U_3O_8 (1973 prices) at 1 000 000 tonnes. This sounds generous, but it must be borne in mind that without the use of breeder reactors, about 1 800 000 tonnes are expected to be needed during the period to 1990, assuming that predictions of increased demand are accurate. This is worrying in the extreme. Since discovery rates have been running at only 65 000 tonnes of uranium per year, it is clear that a major increase in discovery rate is needed. The situation changes dramatically if breeder reactors (Chapter 6) are introduced, for it is then possible to use all the available uranium as fuel, instead of only a small proportion. With breeding, about 8×10^{10} joules can be obtained from each *gram* of uranium, so that 800 000 tonnes (the amount available at $ 10 per pound) represents about 10^{23} joules. This is comparable with the coal resources, and is thus attractively large. It gets more dramatic, however, when it is realised that not only the 'cheap' uranium, but also the more expensive sources, are worth exploiting because of the extremely high energy return. For example, it has been estimated that the Chattanooga black shale, a very low-grade uranium-bearing ore with only 60 grams of uranium per tonne of shale, has an energy equivalent to 1000 tonnes of coal per square metre of land surface.

Uranium used in breeder reactors is thus a very important potential source of energy. It is dwarfed, however, by another source of nuclear power, fusion. As we shall see in Chapter 6, there are two likely candidates for useful fusion reactions, the tritium-deuterium reaction and the deuterium–deuterium reaction. The available resources in the Li–D case are limited by world supplies of lithium to about 2×10^{23} joules, which is again an attractively large figure. If the D–D reaction succeeds, however, it has been estimated that enough energy would become available to keep us going at the present rate for 6 million years!

Nuclear power thus represents a major factor in world fuel policy. If breeder reactors can be made acceptable, the so-called energy crisis will go away for a very long time. Against this is the major worry over the safety of such devices. If the fusion reactor based on the D–D reaction can be made to work, the energy crisis will vanish for ever. No doubt there would be other problems but the prospect is quite exciting.

The immediate conclusion to be drawn from the foregoing is that there is indeed no energy crisis at all. There is rather a short-term problem of Middle Eastern oil supplies, and an immediate longer term problem of diminishing oil reserves. Similar problems, though much less immediate, apply to coal. Vast supplies of other sources of energy exist; the only thing lacking is the technology to render them useful to man in an acceptable and safe manner.

3 Energy Conversion

Central to any discussion of energy, and indeed central to the whole of science, is the tenet that energy can be converted from one form to another according to well-defined laws. This convertibility has occasionally been obscured by the wide range of different units (see Appendix) which has been used in the past to express different types of energy (e.g., Btu for heat, calories for food, kWh for electricity). But with the adoption of a unified system for all forms of energy, in which the watt is the basic unit of power, and the joule (or sometimes kWh) is the basic unit of energy, this obstacle has been removed, at least in most cases. In this chapter we consider the basic laws of energy conversion, and then go on to look at examples of conversion processes.

The Laws of Thermodynamics

Under the rather forbidding title of 'The First Law of Thermodynamics' lies the simple concept that energy can neither be created nor destroyed, but only converted from one form to another. For example, the energy which the sun emits in such prodigious quantities does not arise spontaneously but, instead, is thought to arise from the thermonuclear reactions in which light atomic nuclei are combined to form heavier nuclei with the emission of energy, mainly in the form of heat and light. (Exactly how the energy content of the sun got there in the first place is a matter for theological rather than scientific speculation!) A more earthly example is that of a motor car driving uphill at a steady speed. Chemical energy in the fuel is converted by the engine into heat and thence into mechanical energy, turning the wheels and driving the car upwards against the force of gravity, giving the car potential energy as it does so, This is not, of course, the whole story, as can be seen when the car, instead of climbing a hill, drives along a level road. If the engine is stopped and the gearbox put into neutral, the car slows down and eventually stops, which appears to be a total loss of energy (kinetic energy in this case). However, accurate accounting reveals that

(1) the tyres are warmed during the motion of the car, representing a conversion of the kinetic energy of the whole car to heat in the tyres;
(2) motion of the car through the air generates turbulence in the air, which is a transfer of kinetic energy from the car to the air, and eventually the air turbulence is dissipated as heat;

(3) small losses occur due to friction in bearings, maladjusted brakes and so on, all of which end up as heat.

From the above it is evident that the ultimate fate of all forms of energy seems to be conversion into heat. There is another aspect as exemplified by the car engine. Fuel is burnt at a high temperature; some of the heat is converted into work, and the remainder, the waste heat, is rejected at a lower temperature. Both of these observations are different aspects of the Second Law of Thermodynamics, a law which can be stated in numerous ways, most of them highly philosophical or mathematical and thus of little use to the present discussion. One simple expression might be 'Heat always flows from hotter bodies to colder ones, and cannot flow in the opposite direction', a fact which appears to be trivial and obvious. However, mathematical manipulation of this trivial point yields a much more significant piece of information: 'It is impossible to convert heat completely into work.'

We are now in a position to deduce some very important practical limitations on energy conversion. Any device which involves burning fuel in order to produce mechanical power must be less than 100% efficient. However, devices which convert fuel into heat, or which avoid the use of heat at all, can be, in principle, 100% efficient. This deduction is illustrated in Table 3.1,

Table 3.1 Efficiencies of energy converters

Large boiler	90%	Heat → heat
Domestic boiler (gas)	75%	
Domestic boiler (oil)	70%	
Domestic boiler (coal)	60%	
Hydro-electric turbine	90%	Mechanical ⇄ electrical
Large electric motor	90%	
Large electric generator	90%	
Small electric motor	70%	
Steam turbine	45%	Heat → mechanical (Carnot limited)
Diesel engine	40%	
Car engine	25%	
Steam locomotive	10%	

(These are typical upper limits. In practice most devices have lower efficiencies than those quoted. This applies especially to domestic installations where maintenance is often done badly or neglected altogether!)

in which the efficiencies of various types of energy converters are compared. It is clear that the best efficiencies are achieved by machines which do not burn fuel, but that devices which convert fuel to heat are not much worse. The lowest efficiencies are in all cases achieved by *heat engines,* which convert the energy of fuel into heat and thence into mechanical energy. In all these cases, over half of the energy of the fuel is being wasted as heat.

Why is this so? The explanation involves the natures of heat and of mechanical energy.

In 1824, the French engineer Sadi Carnot published a book entitled 'Reflections on the Motive Power of Fire', in which he analysed the way in which a heat engine worked. He came to the conclusion that if an engine had a boiler temperature of T_H and a cooling water temperature of T_C, both T_H and T_C being absolute temperatures, the maximum efficiency that could be achieved was $(T_H - T_C)/T_H$. This was a theoretical maximum, and any practical engine must have an efficiency less than this. At the time the theory of Carnot could be understood only by using rather obscure mathematical and philosophical reasoning, but with the aid of modern atomic theory it is possible to explain it in a very graphic way. We now know that heat is the manifestation of the kinetic energy of the atoms in a

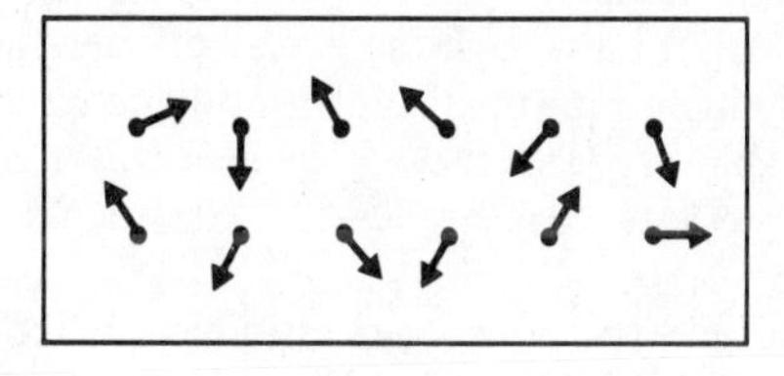

Fig. 3.1 Schematic representation of the motion of the molecules of a substance due to thermal energy.

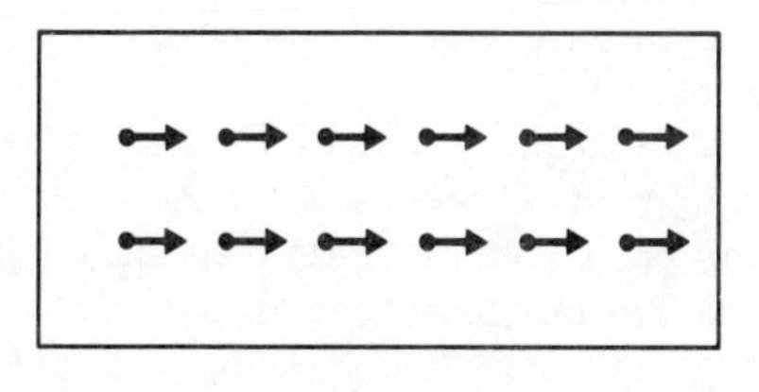

Fig. 3.2 Schematic representation of the motion of the molecules of a substance due to kinetic energy (motion of the whole body).

substance (Fig. 3.1). We also know that when a body moves, all its constituent atoms move also, so that mechanical energy, like heat, is a manifestation of the movement of the atoms (Fig. 3.2). Yet there is a difference – the motion of the atoms in mechanical energy is all in the same direction, that is, it is directed, while the motion in heat is in all sorts of different directions, which add up on average to zero, that is, the motion is random.

To convert heat into mechanical energy, it is necessary to convert random motion into directed motion, which is rather like asking a shuffled pack of cards to arrange themselves spontaneously into their respective suits when the pack is dropped. A mathematical analysis of this situation shows that it can only be achieved at the cost of loss of efficiency. By contrast the converse process, conversion of mechanical energy into heat (as in the braking system of a car) can be done with a theoretical efficiency of 100%, which is not surprising when it is remembered that this is equivalent to asking a carefully arranged pack of cards to rearrange itself into a state of chaos when dropped on the floor – a situation which may well be familiar to less than honest card players.

Efficiency of Electricity Generation

The most publicised example of the inefficiency of the heat engine is in electricity generation, for although the overall efficiency of modern electric plant is remarkably high in comparison with other sources of power such as the petrol engine, the very large scale of electric power generation leads to a feeling that it represents extravagance on a gigantic scale. How far is this criticism justified?

The efficiency of electric power generation has increased dramatically over the years. Around 1900 the average was around 5%, so that about 95% of the energy of the fuel was wasted as heat, either up the flue, or into the cooling water. Today the efficiency, averaged over both old plant and new, is near 30%, and the newest plants have reached over 40%. Clearly this represents a great improvement. Nevertheless, even in the latest stations, about 60% of the energy in the fuel is being wasted. This is simply because at present there is no alternative to the heat engine as a means of large scale electricity generation.

What can be done about this problem? The short answer is nothing, and yet this is an overstatement. As mentioned at the start of this section, the efficiency of modern plant is vastly better than that of older equipment. This has been achieved mainly by a steady rise in the temperature of the boilers, and a consequent rise in their pressure. It is easy to see that if T_H in Carnot's equation is increased, while T_C is kept constant, the theoretical efficiency rises. But there is a limit to the extent to which this can be done, for the boiler material must be able to withstand the very high temperature and pressure, and do so completely reliably for years. Modern boilers are already operating at such high temperatures and pressures that it seems rather unlikely that much further improvement can be expected.

There are ways in which the 'high temperature' heat of the fuel can be used to some extent before it gets to the boiler. These include gas turbines and magnetohydrodynamics, both of which may make it possible to raise the effective T_H without necessitating exotic boiler materials. Unfortunately there are practical problems, and it seems unlikely that techniques such as these will make any substantial contribution to efficiency in the near future.

A glance at the Carnot equation shows that it is possible to raise the efficiency, not only by raising T_H, but also by lowering T_C, that is, by lowering the temperature of the cooling water. Unfortunately there is no convenient large source of coldness which is colder than the sea or a large river, so improvement in this direction seems unlikely.

A final, and perhaps rather drastic, proposal for increased efficiency of electricity generation is to abandon the heat engine altogether, and to use a method of converting the energy of a fuel directly into electricity. This is indeed achievable already, in the shape of the fuel cell. Fuel cells are still in the experimental stage, and although they are used for certain purposes such as space flight where their advantages are great and where cost is of no importance, the immediate prospects for large-scale power generation are poor.

An alternative approach to the problem of the waste heat is to consider ways in which it can be used for useful purposes. The problem here is that to achieve the best efficiency in the generation of electricity, it is necessary to have the lowest possible temperature in the cooling water, and such low temperature heat is of little use. There have been several suggestions for activities such as fish-farming in the warm waters near the outlets of a power station, and many of these schemes appear practical on paper, but whether they can be used in practice has yet to be established. Another, more radical, proposal is to increase the temperature of the cooling water, and so reduce the efficiency of electricity generation, in the interests of using the waste heat usefully, for example, for domestic space heating. Such district heating schemes are common in many countries, notably Scandinavia.

Much has been written recently about the supposed waste of resources that occurs when electricity is used for purposes in which the direct use of fuel might appear to be more efficient. Typical examples have been domestic space and water heating, which if done by means of an oil-fired boiler have an efficiency of perhaps 60%, while if done with electricity can only achieve perhaps 30% because of the inefficiency of the power station. This is true, if efficiency in the sole criterion. However, there are other considerations. The typical oil-fired boiler burns either kerosene, which can be used for aircraft fuel, or gas oil, which can be used for heavy road vehicles and railways. In other words, it uses a very useful high-grade fuel. In contrast, an electricity station uses either low-grade coal or heavy residual oil, or even nuclear or hydro-power, none of which is much use for anything other than such very large scale static processes. Which do we choose, the efficient use of a fuel which is essential for other purposes, or the inefficient use of a fuel which is useless for other purposes? In the world of practical economics such decisions are made on the basis of the costs of the different alternatives, and while that may not be the ideal criterion, it is preferable to a decision made solely on a consideration of relative efficiencies.

Transmission of Energy

The most straightforward means of conveying energy to its consumer is simply by transport of the primary fuel. Of more interest to us at present is the conversion of the energy of the primary fuel into some other form which is in some way more convenient to transport or to use. The most obvious and important example of transmission of energy is electricity, but there are other methods which are already of importance or may become so.

Electricity is entirely a secondary source of energy, so our first consideration must be the source of primary energy from which it is generated. The usual source is the steam turbine (Fig. 3.3), the operation of which can be

Fig. 3.3 A large modern steam turbine installation. (Reproduced by permission of GEC Turbine Generators Ltd.)

summarised as follows. Fuel is burnt, the heat liberated raises steam at a high pressure and temperature, this rotates a turbine (losing energy by reduction of both pressure and temperature), and enters a condenser where it becomes liquid water which is pumped back to the boiler. The steam turbine is a heat engine, and the ultimate limit to its efficiency is given by the Carnot equation

$$\eta = \frac{T_H - T_C}{T_H}$$

Putting in typical figures, the maximum theoretical efficiency is about 60%. In practice the efficiency is usually about 45% or less.

The water in the boiler is of very high purity, to prevent corrosion and the deposition of 'fur'. The pressure is raised well above atmospheric to elevate the boiling point of the water; typical figures for a modern power station are 150 atmospheres and a temperature of 550°C. The boiler is constructed of steel tubes placed parallel to each other all round the walls of the combustion zone. The water passes through these tubes, so that the burning fuel is effectively surrounded by a wall of water, providing very efficient heat transfer. As the water boils, the steam passes through another set of tubes known as the superheater. The object of this is to heat the steam to as high a temperature as possible. This is both to improve the efficiency of generation and to prevent condensation into water droplets. (Although we have distinguished between water and steam in this discussion, it is often impossible in modern boilers to tell where the dividing line is between the two. This is because at very high pressures and temperatures there is actually no difference! The water, or steam, is said to be 'super-critical'.)

In the turbine, the steam is made to expand against a series of blades whose shape and angle are designed to extract as much energy as possible. During this process both the temperature and the pressure fall, but the pressure drop is more significant and so the volume of the steam increases. For this reason the later stages of the turbine have to be larger than the earlier ones, and it becomes difficult to combine all the stages into one assembly. It is usual to have two or even three distinct turbine assemblies working over different pressure ranges. They are often on the same shaft, and rotate at the same speed, but even this may cause problems because the large diameter of the low-pressure turbine, combined with a high speed of rotation, creates a large central force on the blades. This problem can be circumvented by running the low-pressure turbine as a separate unit at a lower speed.

From the low-pressure turbine the steam passes to the condenser, where it condenses to water and is returned to the boiler. The condenser is cooled by a flow of ordinary unpurified water, which is often obtained from the sea or a large river. In the case of inland power stations, where river water may not be available in large enough quantity, an alternative arrangement may be used. The cooling water is circulated to a cooling tower, in which it is sprayed from fine nozzles into a rising stream of air. Some of the water evaporates, cooling the rest, and the cooled water is recirculated to the condensers, while the evaporation loss is made up from a river. This arrangement makes more efficient use of the cooling water than does direct circulation from and back to a river, but it has its own problems, principally that it raises the humidity of the surrounding air and can cause mists in unfavourable weather conditions. Dry cooling towers, in which the cooling water does not evaporate but passes through a 'honeycomb' rather like a car radiator, have been tried on an experimental basis, but the cost of construction is inevitably higher than that of conventional towers, and it seems unlikely that they will be used except where unavoidable.

The turbine drives an alternator (an alternating current generator) which converts the mechanical energy of the turbine into electrical energy. The use of alternating current (a.c.) arises from the great ease with which a.c. can be changed in voltage. High voltages can be transmitted over long distances very efficiently and economically, but lower voltages are more convenient for both the generator and the consumer. Changing an a.c. voltage is done by a transformer, which is an extremely efficient (99%) and reliable device. In contrast, changing the voltage of direct current (d.c.) requires a mechanical or electronic device which is much more expensive and rather less efficient. Nevertheless, the use of d.c. is important in certain special circumstances.

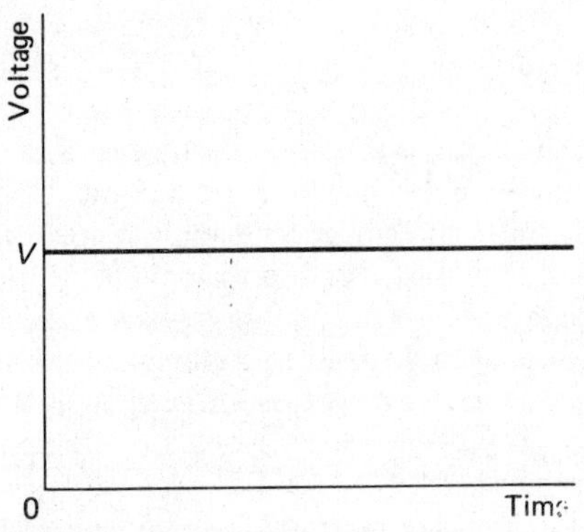

Fig. 3.4 Direct current (d.c.) electricity.

A d.c. supply, such as that from a battery, has a constant voltage, so a graph of voltage against time is simply a straight line (Fig. 3.4). An alternating current supply has a voltage which changes constantly in a wave-like manner, as shown in Fig. 3.5, and this behaviour is known as a sine wave. The voltage increases from zero to a maximum value V_{max} in a positive direction, then decreases to zero and becomes negative, reaching a maximum negative value of V_{max} before returning to zero. The entire cycle then repeats itself at a regular frequency; in the mains electricity supply of Europe it repeats every 1/50th of a second, and in the USA it repeats every 1/60th second. As shown in Fig. 3.5, the oscillating voltage may be regarded as the projection of a rotating arrow onto the voltage axis.

The average voltage of an a.c. supply is obviously zero, since it spends as much time negative as it does positive, and yet it clearly behaves differently from a 'supply' at a steady voltage of zero. Consequently a different type of average, the root-mean square (r.m.s.) voltage, is needed. There is a

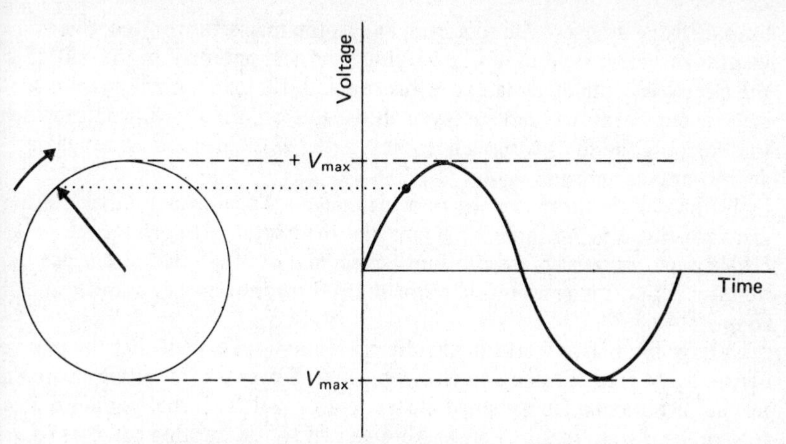

Fig. 3.5 Alternating current (a.c.) electricity. Note how a wave results from the projection of the rotating arrow.

simple relationship between the r.m.s. voltage V and the maximum voltage V_{max} referred to earlier. The relationship is

$$V_{max} = \sqrt{2}V = 1.414V$$

so that the maximum voltage of a 240 volt a.c. supply is 339 volts.

The simplest possible alternator is the bicycle lamp generator, which is sketched in Fig. 3.6. A magnet is rotated on a shaft between the ends of a piece of soft iron, on which is wound a coil of copper wire. The magnetic

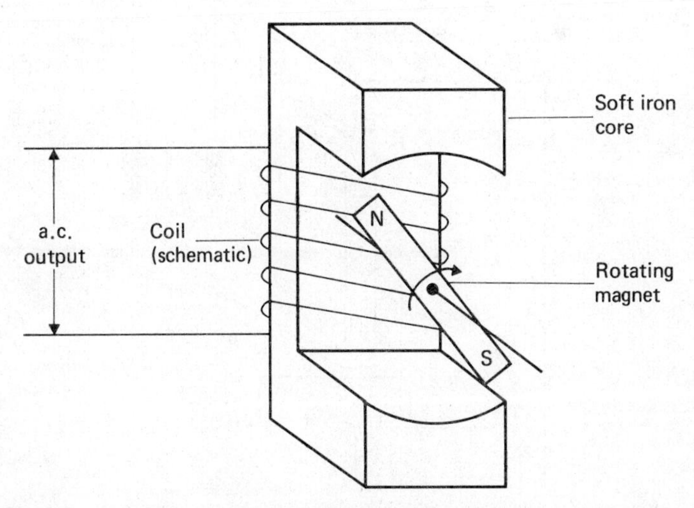

Fig. 3.6 A simple a.c. generator (alternator) with permanent magnet rotor.

lines of force intersect the soft iron, and as the magnet is rotated, the varying magnetic field induces a varying electrical potential in the coil. If the circuit is completed via two wires and a useful load such as a lamp, an electric current flows, and energy is absorbed from the rotating magnet, so that keeping the shaft turning becomes harder work. This energy appears at the lamp as heat and light.

In this simple alternator the rotating magnet is analogous to the rotating arrow of Fig. 3.5. A simple 2-pole magnet in an alternator producing a.c. at 50 Hz must rotate at 50 revolutions per second or 3000 revolutions per minute; a 4-pole magnet must rotate at 1500 revolutions per minute, and so on.

A large alternator works in exactly the same way, except that the permanent magnet is replaced by an electromagnet. Power to excite this electromagnet is obtained from a small d.c. generator which is usually mounted at the end of the alternator shaft and connected to the rotating coil by slip-rings (Fig. 3.7).

A large alternator is exceptionally efficient (about 99%), but because of the very large amount of power handled even 1% loss is significant. A typical large alternator may be rated at 500 MW, and 1% of this is 5 MW or 5000 kW! This power loss appears as heat, and the removal of such a large

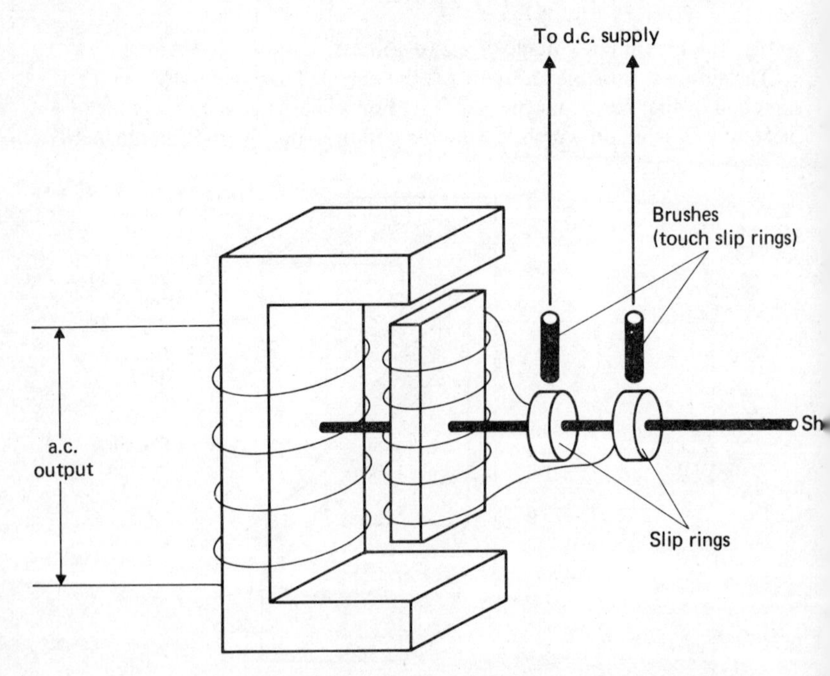

Fig. 3.7 Principle of a large a.c. generator with d.c. excited rotor.

amount of heat cannot be left to natural convection. It is usual to use hydrogen gas as a coolant.

Transmission voltages depend on the distance over which the power is to be sent. Typical voltages in the United Kingdom are 400 kV, 275 kV, 132 kV and 66 kV. There is a tendency in the USA to use higher voltages, partly because the distances involved are greater.

In overhead power lines air is used as an insulator where possible but at the pylons (also known as towers) a string of glass or ceramic insulators is needed, and this string is longer for higher voltages. The lack of insulation on the conductors themselves would appear to present a safety hazard to birds, but this is not so. Birds are perfectly safe provided they touch only one conductor, for then no current can flow through them. The separation between the conductors is made large enough to present no hazard to even the largest bird. Aircraft are a more difficult problem, but it is most unusual for aircraft to fly as low as the tops of transmission lines except near airfields, where the power lines are put underground to avoid risk. The falling of a live wire would be dangerous, however, and to prevent accident the transmission system is protected by automatic equipment which switches off the line as soon as a break is detected. This protection system works so rapidly that the line is switched off before the falling conductor reaches the ground, but as a second line of defence, another protection system can detect the earthing of a conductor and once again switch the line off.

Distribution of electricity to consumers is at a lower voltage, typically 33 kV or 11 kV. Again the lines are overhead where possible, but in urban areas they are often put underground for aesthetic reasons.

The most noticeable feature of the distribution and transmission lines is the arrangement of conductors in threes. To understand the reason for this it is necessary to look at the nature of alternating current in more detail.

Earlier we showed that alternating current could be described as a sine wave generated by projecting a rotating radius on to the diameter of a circle, and as we saw, two wires were needed to connect the alternator to the load. Let us now see what happens when we consider three radii, at 120° to each other. (Fig. 3.8a). Imagine each radius to be the magnet of a separate alternating current in a coil. Normally each alternator coil will be connected to two wires, requiring six in all. But if the coils are connected as in Fig. 3.8b, only four wires are needed, since the fourth wire acts as a return conductor for all three coils. The three wires connected to the outer ends of the coils are said to carry the three *phases*, and the fourth wire (which is usually connected to earth) is known as the neutral. The saving of two wires in a long transmission line is a considerable economy. Further, if the three wires all carry identical currents, and are out of step with each other by 120°, it is easy to show by vector algebra that the current in the neutral conductor is always zero. In this situation the neutral conductor is unnecessary and can be removed, leaving only three conductors instead of the original six,

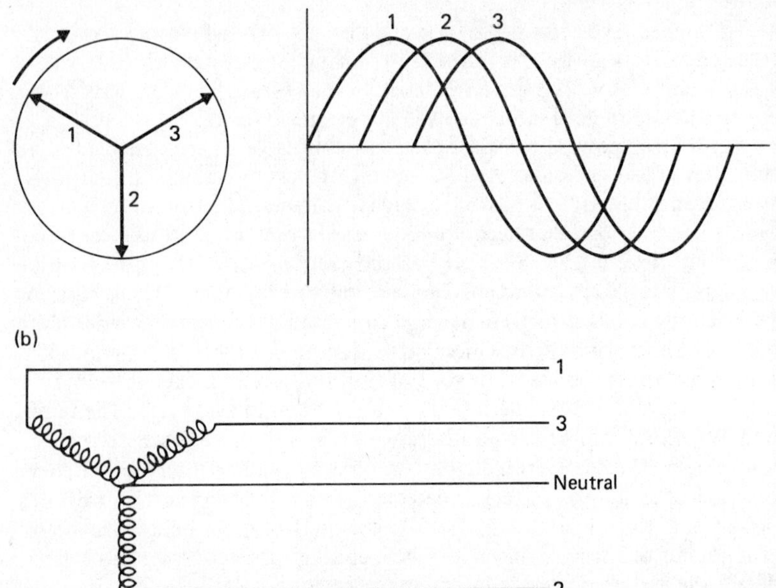

Fig. 3.8 Three-phase a.c. (a) Principle (b) 'Star' connection eliminates two wires.

which is an even greater economy. It is mainly for this reason that three-phase a.c. is almost always used in electricity transmission.

One further advantage of three-phase supplies needs to be mentioned. A three-phase electric motor is rather simpler to design and construct than an equivalent single-phase motor, and it is capable of running more smoothly. For this reason it is usual to supply three-phase electricity to consumers, such as workshops and garages, who have large numbers of electric motors.

Despite the advantages of a.c. transmission, there are several problems. The most obvious of these is that although the r.m.s. voltage V accurately describes its 'useful' voltage, the maximum voltage V_{max} occurs twice every cycle, and this must be allowed for when insulating the wires carrying the supply. For ordinary domestic mains the effect is small, since it is little more difficult to insulate for a maximum of 339 volts than it is for the r.m.s. value of 240 volts. For high-voltage transmission, however, the extra voltage is a problem because it necessitates longer strings of insulators and higher pylons, both of which add to the cost of the line. For undergound cables the situation is even worse, because the whole of each wire has to be insulated to withstand the extra voltage. For this reason it has been found desirable in some circumstances to convert the a.c. into high-voltage d.c. for transmission, and to convert it back to a.c. at the other end of the line. The conversion is accomplished by very large mercury-arc discharge tubes, or sometimes by equivalent devices in solid-state electronics. Such conversion devices are very expensive, and can only be justified when the transmission line is very long, as can happen in the USA, or where a long section of the line has to be underground or underwater, as in the connection between England and France.

Another problem of a.c. networks is also alleviated by the use of d.c. interconnections. In the whole of an interconnected a.c. system all the alternators supplying power must run at exactly the same speed. When a new alternator is to be brought in, it must first be run up to speed (e.g. 3000 revs/minute) and then synchronised with the existing mains supply so that its waves are in step with those of the mains. Only then can it be connected. Once connected, the power to the turbine is increased and the wave generated by the alternator starts to 'lead' the wave in the mains by a few degrees, and in this condition power is fed from the alternator into the mains. But despite the 'lead' the speed of the alternator remains 3000 revs/minute.

The same situation exists when an attempt is made to interconnect two existing networks by means of a new transmission line. Before the switch can be closed, both *networks* must be running at the same speed and must be synchronised. Speeding up and synchronising an entire network is a much more difficult problem than doing so for one alternator. If the interconnection is made by a d.c. transmission line, however, both networks can run at their own speed. Only the conversion devices need to be synchronised, and this can be done electronically.

Energy Storage

A final important aspect of energy conversion is storage. Energy can be stored in various ways, for example, chemically as in a fuel such as crude oil or in a processed fuel such as petrol, electrochemically as in batteries, or mechanically as in the potential energy of water behind a dam. Energy storage will clearly be necessary whenever the primary source of energy is only available on an intermittent basis, as is the case with solar power, or when mobility of the power source is of prime importance, as in transport.

As an example, consider the use of storage in conjunction with a wind power scheme. The wind power can be used to pump water uphill, where it is a potential source of electrical energy, via a turbine. Alternatively, it can be converted directly to electricity which can then be further converted into electrochemical energy stored in batteries, into thermal energy stored in a hot water tank, or into the chemical energy of hydrogen and oxygen produced by the electrolysis of water.

The most important example of the need for energy storage is probably in the large-scale generation of electricity. The generating boards have the job of providing sufficient power to meet demand. This would be a relatively simple engineering problem were it not for the fluctuations in consumption

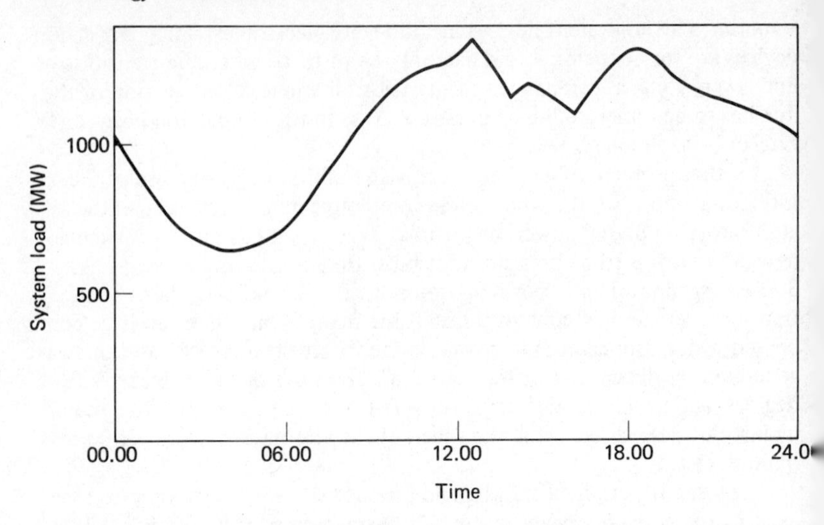

Fig. 3.9 A typical winter daily load curve for an electricity network (by permission of the Electricity Supply Board of Ireland).

during a given day. This is illustrated in Fig. 3.9. There must be sufficient power available to meet peak demand and, in an ideal world, production and demand should be closely matched at all times. But it is no simple matter to shut down or restart large fossil fuel or nuclear power stations, and even to vary their outputs degrades their efficiency. Consequently, it is in the interests of the generating boards to run their large efficient power stations more or less continuously and to store against future requirements the excess power which they produce when demand is low.

The boards have attempted to overcome demand fluctuations in two ways. The first of these involves no direct energy storage at the production end but instead strives to achieve a smoothing of the load curve by offering reduced 'off-peak' tariffs when demand is low, for example, during the night. This 'off-peak' power can be used for space heating provided a suitable means of thermal storage is available to the consumer. In the United Kingdom it has been customary to use high density alumino-silicate bricks in storage heaters whereas in Eastern European countries iron storage heaters are favoured. The disadvantage of all such heaters is their long response time which can lead to much electrical power being wasted on warm winter days. Underfloor heating which uses the concrete floor of a building as the storage medium is becoming increasingly common. Although this suffers the same sluggishness as storage heater systems, the larger mass and heat-emitting surface of the concrete floor means that low temperature storage, which is less likely to cause draughts, can be employed. Underfloor heating also leads to a smaller and more uniform vertical temperature gradient in a room which would appear to be more comfortable for the occupants.

Pumped Storage

As an alternative to this the generating boards can store energy prior to transmission to the consumer by the use of *pumped storage.* Here the surplus electrical power plays an analgous role to that of solar energy in hydroelectric schemes. The surplus power is converted into potential energy by pumping water into an upper reservoir in which it is stored. When demand is high, the sluice gates of the upper reservoir are opened and the water is allowed to run down into a lower reservoir via a set of hydraulic turbines. In reality the pumps and the turbines are one and the same, with different modes of operation corresponding to different senses of rotation of the turbine blades. In a modern pumped storage scheme as little as 20% of the electrical energy is lost in one complete pump–generate cycle.

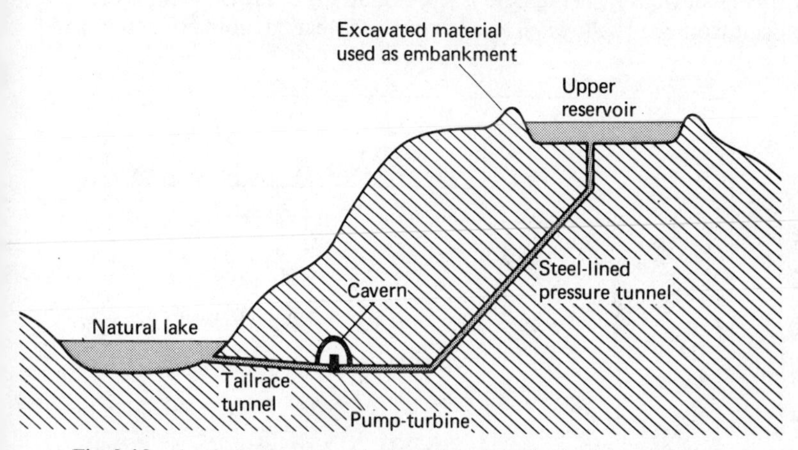

Fig. 3.10 A schematic representation of a pumped storage installation.

The layout of a typical installation is illustrated in Fig. 3.10. A natural lake in hilly countryside is an ideal candidate for the lower reservoir while the top of a nearby hill can be excavated to form the upper reservoir. The flowpath for the water can either be in the form of pipelines drilled through the hillside as in the diagram, or pipelines running down the hillside.

As an indication of the scale of pumped storage schemes, it is worth noting the specifications for some recently completed and some currently planned projects. The Turlough Hill installation of the Electricity Supply Board of Ireland has an output capacity of 292 MW with a maximum stored energy per cycle of 1860 MWh. These figures are dwarfed by the Central Electricity Generating Board Dinorwic scheme in Wales which has an 1800

MW input and a storage capacity of 7.8 GWh per cycle. An even larger installation has been devised by the Consolidated Edison Company of New York for Storm King on the Hudson River. It will have an initial power capacity of 2000 MW. By 1971 the United States had more than 7000 MW of pumped storage in operation or being installed, with planning permission pending for a further 10 000 MW.

It is clear then that the electricity producers have considerable faith in pumped storage and the reasons are not difficult to find. In the first place it can contribute to peak load generating capacity. Secondly, the turbines can be kept in a state of readiness during the day by running them at, say, 10% of peak capacity. This allows them to be kept in synchronisation with the rest of the electricity network and to respond within a few seconds of a major failure elsewhere. Similarly, when pumping is taking place, the turbines are an effective load which can be shed rapidly in the event of a sudden upsurge in demand. Thirdly, pumped storage can be used to smooth the daily load curve (the effect of the Turlough Hill scheme on the ESB load curve is obvious from Fig. 3.11). This smoothing of the curve releases steam plant from load following, so allowing it to be used more efficiently and

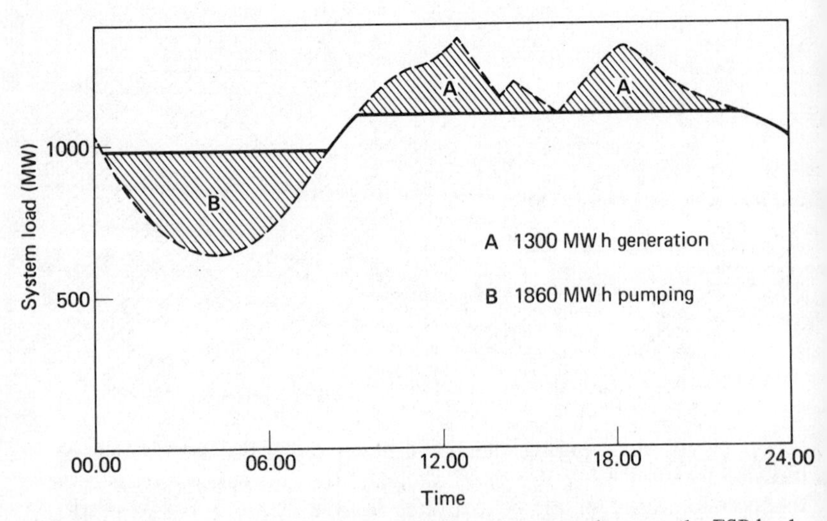

Fig. 3.11 The effect of the Turlough Hill pumped storage scheme on the ESB load curve (by permission of the Electricity Supply Board of Ireland).

raising the overall efficiency of electricity production. It must be remembered though that pumped storage rarely saves fuel as such. Losses incurred at the pumped storage station usually offset the savings which accrue from the use of high efficiency thermal stations by night and the displacement of inefficient plant by day.

Hydrogen Economy

It is conceivable that the present trend towards an electricity dominated energy scenario will be accelerated in the future, especially if the nuclear reactor programme continues to be given high priority. As the fossil fuels become scarce and more expensive, or as their use becomes more and more confined to the petrochemicals industry, then there will be a demand for a fuel to replace the petroleum derivatives which are presently used in the various modes of transportation. One candidate for this role is hydrogen and the conversion and storage of energy as hydrogen forms the basis of the futuristic 'hydrogen economy'.

Hydrogen's attraction as a fuel stems from its abundance in ordinary water. At present most hydrogen is produced from petroleum feedstock but there are some electrolytic plants in operation. Unfortunately, the efficiencies of large-scale electrolytic cells are at best 65–70% and it is estimated that plants producing more than 7000 tonnes per annum would cost twice as much per unit energy content produced as the electricity consumed. Clearly, then, conventional electrolysis produces expensive hydrogen, and considerable improvements in cell design will be necessary before cheap hydrogen is available. As an alternative production method, the thermal dissociation of water by the waste heat from nuclear reactors has been suggested. Unfortunately, the dissociation temperature of the water molecule is 2500°C, but it is possible to split water by using multi-step chemical processes operating below 1000°C. One possibility identified by the Euratom research organisation is a three step process involving ferrous chloride:

$$6FeCl_2 + 8H_2O \rightarrow 2Fe_3O_4 + 12HCl + 2H_2 \quad (650°C)$$

$$2Fe_3O_4 + 3Cl_2 + 12HCl \rightarrow 6FeCl_3 + 6H_2O + O_2 \quad (200°C)$$

$$6FeCl_3 \rightarrow 6FeCl_2 + 3Cl_2$$

$$2H_2O \rightarrow 2H_2 + O_2$$

Temperature control is, however, of paramount importance. If the first reaction is allowed to run at only a slightly lower temperature, ferrous hydroxide is produced and no hydrogen.

As well as its abundance, hydrogen also has the advantage of being an attractive fuel to use. It is at least comparable with gasoline in its specific energy content. Its energy content per unit weight is twice that of octane, although on a volume basis (as a cryogenic liquid) hydrogen's energy content is nearly three times poorer than octane. As it burns with an exceptionally high flame temperature (about 2500°C) it is an ideal candidate for use in heat engines. Moreover, it burns cleanly to produce water. (Some traces of nitrogen oxides are formed from the air near the flame but even these can be removed by catalytic combustion at 100°C.)

Once manufactured, the distribution and storage of hydrogen should not present insuperable problems. Pipeline distribution of natural gas is already common in many countries and the same or similar pipelines could equally well be used for hydrogen. Interestingly enough, the same pipeline could transmit the same power equivalent (energy content per unit time). Although hydrogen has approximately one-third the energy content per unit volume of natural gas, its lower density and viscosity allow the pipeline to handle three times the flow rate, the only difference being a slightly higher pumping power for hydrogen. Storage of hydrogen in cryogenic tanks is already a well established technique, and 5000 m^3 tanks of liquid hydrogen have already been built by NASA for use in the space programme. Of course, there is always a risk of explosion when hydrogen comes into contact with oxygen or air but, with sensible precautions, hydrogen, whether as a gas or as a liquid, can be handled as safely as gasoline. A further improvement in storage safety may be brought about by the use of metal hydride 'sponges'. Thus magnesium, for example, will absorb more than 1000 times its own volume of hydrogen and release it on warming. Indeed, a magnesium hydride fuel tank would only be four times as large as a gasoline tank for the same energy content. Unfortunately, at atmospheric pressure magnesium hydride does not dissociate below 287°C, so a hydride fuel tank must await the discovery of a hydride with a capacity similar to that of magnesium hydride but with a much lower operating temperature.

Hydrogen has many potential uses. It is already used as a feedstock in the fertiliser industry (ammonia synthesis is the major use of hydrogen today), and, if fossil fuels were to become scarce, it could become the feedstock for what we now term the petrochemical industry. Because of its cleanness it is an ideal fuel for domestic cooking and heating requirements. It needs no flue and would, therefore, be more efficient than conventional fuels, and even provides its own built-in humidification system! It is also an ideal standby fuel for gas turbine electricity generators or could be used for local power generation in isolated areas. Finally, it is an ideal fuel for transportation. A conventional internal combustion engine will run on hydrogen provided some modifications are made to the carburettor to prevent pre-ignition. It could also replace kerosene as an aircraft fuel, an application in which its lightness would be a positive advantage.

It is worth noting that, as in so many other technological advances, the hydrogen economy and the thermonuclear reactor were both foreseen by that pioneer science fiction writer Jules Verne. In 'The Mysterious Island' he writes:

> '. . . and what will men burn when there is no coal? Water, Yes, my friends, I believe that one day water will be employed as a fuel, that hydrogen and oxygen which constitute it, used singly or together, will furnish an inexhaustible source of heat and light.'

Batteries and Fuel Cells

Of course, hydrogen is not the only mobile source of energy available for the transport of the future. Electric cars, in the form of milk delivery vans and fork lift trucks, are already a familiar sight in the United Kingdom and elsewhere. Could the battery powered vehicle become the transport of the future? The answer must lie in the technological development of batteries and their related electrochemical power sources, fuel cells.

Batteries are of two main types, primary and secondary. In a primary battery chemical energy extracted from the constituents of the battery is converted into electrical energy via reactions in the electrolyte. Once the constituents in their active form have been exhausted the battery is discarded. In a secondary battery, on the other hand, the reactions can be reversed by the input of electrical energy and the battery recharged. This charge–discharge cycle can be repeated many times before the battery comes to the end of its useful life which makes the secondary battery much more important in applications such as electric cars. The familiar lead–acid accumulator (as used in cars) serves to illustrate the mechanism of the secondary battery.

The lead–acid cell comprises two porous lead electrodes impregnated with lead sulphate, which is only slightly soluble. The electrodes are immersed in a sulphuric acid electrolyte which is saturated with lead sulphate. During charging the following reactions occur at the electrodes

$$Pb^{2+} + 2e \rightarrow Pb \qquad \text{(cathode)}$$

$$Pb^{2+} + 2H_2O \rightarrow PbO_2 + 4H^+ + 2e \quad \text{(anode)}$$

For discharge, the anode and the cathode exchange roles and the following reactions occur:

$$PbO_2 + 4H^+ + 2e \rightarrow Pb^{2+} + 2H_2O \quad \text{(cathode)}$$

$$Pb \rightarrow Pb^{2+} + 2e \qquad \text{(anode)}$$

so that the overall cell reaction can be written as

$$Pb + PbO_2 + 4H^+ \underset{\text{charge}}{\overset{\text{discharge}}{\rightleftharpoons}} 2Pb^+ + 2H_2O$$

The lead–acid cell has an open-cell voltage (when fully charged) of 2V – six such cells are linked in series in the common car battery. The available voltage decreases as the amount of power extracted from the battery is increased; moreover, the voltage–current curve depends on the degree of discharge of the battery (Fig. 3.12). The other relevant characteristics of a battery, as far as its importance as a portable power source is concerned, are its energy density and power density. The lead–acid battery has a typical energy density of 15 Wh kg^{-1}. Thus a battery of 10 kg will provide 150 Wh of energy in the course of one cycle. The actual performance will

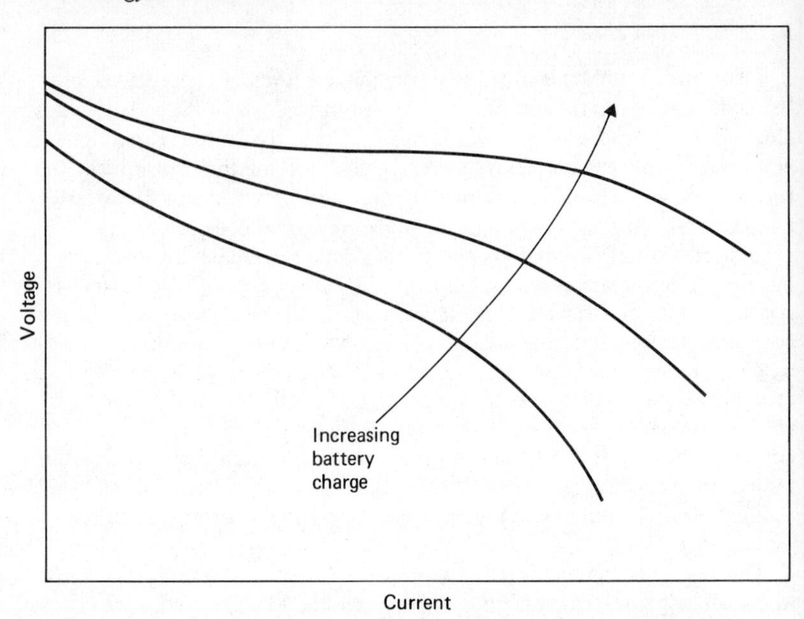

Fig. 3.12 Battery voltage–current curves for different degrees of discharge.

depend markedly on how rapidly the battery is discharged, so this figure can only be regarded as a guide to battery performance under average conditions (say a five hour discharge period). The power density of the battery (about 150 W kg^{-1} for lead–acid) refers to the maximum power available from the battery.

Energy density and power density of an electrochemical power source are the parameters which must be improved significantly if electric cars are to become truly feasible. The energy density limits the range of the vehicle while the power density limits its performance in accelerating and hill climbing. The current state of the art in electric car production is a two-seater packed with batteries, capable of a top speed of 40 mph, and with a range of 40 miles. Even so, it is all but competitive as a commuter car especially as there is no need for the engine to idle when the car is stationary for a short period. (This is the attraction of the electric delivery vehicles which are common in the United Kingdom and which must operate under extreme stop–start conditions.)

What then are the possibilities for further improvement? It would be erroneous to think that the lead–acid battery has reached the end of its development, especially as its storage capacity has been uprated by 35% in the last ten years or so. But for really significant advances one must look to different cell reactions. Nickel–cadmium batteries with sintered nickel electrodes are already available with energy densities as high as 40 wh kg^{-1}

and power densities of 150 W kg^{-1}. However, the limitation to their use is an economic one based on the high cost of nickel and cadmium compared to lead. Silver–zinc batteries, which have high enough energy densities to make them ideal candidates for use in electric cars, also suffer from extreme expense, and in addition from a tendency for the plates to buckle after only a few hundred charge–discharge cycles. Perhaps the best hope for the future lies in the sodium–sulphur battery currently being developed by the Ford Motor Company of America and the Electricity Council's Capenhurst Research Laboratory in the United Kingdom. Strangely enough, it uses a solid electrolyte (β-alumina, which is permeable to sodium ions) and liquid electrodes of sodium and sulphur. This does of course necessitate a high operating temperature of between 250 and 400°C. The energy density of prototype sodium–sulphur batteries is already more than ten times that of a lead–acid battery.

The fuel cell differs slightly in concept from a battery in that a continuous supply of reactants is made available to both electrodes. Fuel is supplied to one electrode and an oxidant, usually oxygen or air, is supplied to the other. Although the first fuel cell was reported by Grove in 1839, development was somewhat pedestrian until the impetus of enthusiasm provided by Bacon. He and his co-workers in Cambridge, United Kingdom, have pioneered the cause of fuel cells since the nineteen thirties. Their work saw fruition with the adoption by NASA of their fuel cells for use in the Gemini space probes. Not only did the hydrogen–oxygen (hydrox) cells provide electrical power, they also contributed a valuable reaction product – drinking water!

The simple hydrox cell is illustrated schematically in Fig. 3.13. Hydrogen gas is passed through a porous nickel electrode where it is encouraged to dissociate into atoms and to adsorb onto the electrode surface as hydrogen ions by the action of a catalyst embedded in the pores. These ions react with hydroxyl ions (which have travelled through the electrolyte from the oxygen electrode) to produce water.

$$H_2 \rightarrow 2H^+ + 2e \text{ (catalyst)}$$

$$2H^+ + 2OH^- \rightarrow 2H_2O$$

At the oxygen electrode, oxygen atoms are adsorbed by catalytic action and are reduced to form hydroxyl ions.

$$O_2 \rightarrow 2\,O \text{ (catalyst)}$$

$$O + H_2O + 2e \rightarrow 2\,OH^-$$

The overall reaction is the production of water, but there is an electron transfer – a current – between the hydrogen electrodes.

The fuel cell is appealing because of its intrinsically high efficiency. Most techniques for generating electricity involve indirect conversion via heat and thereby incur a thermodynamic Carnot limitation. The fuel cell, on the

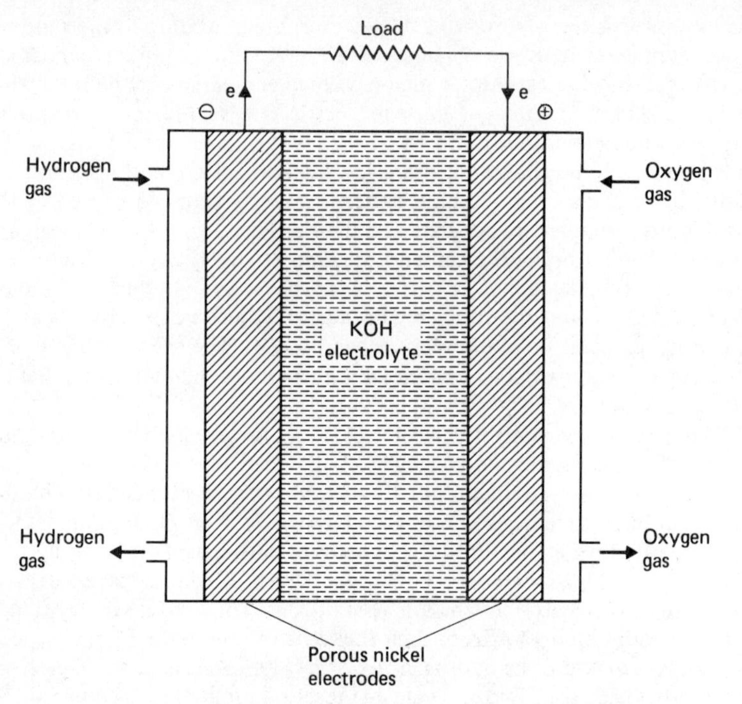

Fig. 3.13 A schematic hydrogen–oxygen (hydrox) fuel cell.

other hand, is a direct conversion device and, in principle, can be 100% efficient. The available energy from a fuel cell is, strictly speaking, a quantity known as the free energy change of the reactants, and amounts to 229 kJ mole^{-1} for the hydrox reaction. To ensure a fair comparison with indirect conversion, we should really compare this figure with the energy obtainable from burning hydrogen in air (the enthalpy change) which amounts to 242 kJ mole^{-1}. Consequently, the theoretical fuel cell efficiency is 95%. In practice other inefficiencies are introduced – imperfections in the electrode structure degrade performance and the electrochemical reaction rate may be limited by the slow diffusion of species between electrodes – and much research is still needed in this area if the full potential of the fuel cell is to be realised.

The most highly developed fuel cell to date is the high temperature hydrox cell constructed by Bacon (Bacon and Fry, 1973). Higher operating temperatures give improvements in performance but there is a corresponding increase in corrosion and a compromise must be reached. Normally, the operating temperature is kept below 200°C. Such fuel cells were used in the historic lunar missions and operated at an overall efficiency of 72%. Their energy density was 880 Wh kg^{-1}, more than two orders of magnitude better than a lead–acid battery. Against this, however, the fuel cell suffers from low power density.

So it seems that we can have high energy density (in fuel cells) or high power density (in batteries) but not both. And yet if we were to employ a hybrid power source combining fuel cells and batteries then we would indeed meet both criteria. This may be the answer for electric cars. The fuel cells would extend the range considerably for steady driving, while the batteries could be switched in when bursts of power are required for overtaking and hill climbing. Moreover, the batteries could be maintained in a charged condition by feeding some of the fuel cells' output to them when the car is cruising.

This short excursion into the study of pumped storage, hydrogen as a fuel, batteries and fuel cells should at least have served to illustrate the importance and diversity of energy storage. Along with electricity generation and distribution it must be rated as an essential feature of energy handling.

Climatic Effects of Power Production

The end result of all energy conversion chains on earth is thermal pollution of the environment. On this tenuous pretext, and notwithstanding our exclusion of pollution from the scope of this volume, it does seem appropriate to consider briefly the effects which man's power production may be having and may have in the future on the finely balanced climate of the globe.

Two points must be conceded immediately. Firstly, the earth's climate has managed to change considerably without any help from *homo sapiens*. The Ice Ages are ample evidence for this. During this century the average temperature of the planet managed to rise by 0.4°C between 1900 and 1945, and has since declined somewhat. Secondly, as we do not begin to understand the complex mechanism behind our weather, and as our models of atmospheric behaviour are manifestly imprecise, it is pertinent to query if we could even detect changes brought about by man. Nevertheless, the point must be made that man's puny activities could well activate the meteorological levers of power and produce perturbations in climate well beyond his expectations. Evidence supporting these fears is afforded by other environmental interference effects such as the denudation and desert formation in Western Australia and the 'Texas Dustbowl' caused by overfarming. Perhaps a simple example will suffice to illustrate the potential for positive feedback in the atmospheric system. Suppose that air pollution leads to an increase in atmospheric absorption and hence to a decrease in average temperature. This would encourage an extension of the polar ice caps. Now, ice has a higher reflectivity than water, so the extension of the ice caps leads to an even further decrease in average temperature and so on. Conversely, an increase in average temperature brought about by man's energy production and consumption could lead to an increased water vapour content in the atmosphere. This could enhance the greenhouse effect (Chapter 2) and thus lead to an even further increase in temperature. Coupled with this positive feedback mechanism, and adding to it, would be that due to melting of the ice caps. Life is not simple for a climatologist!

What then is the direct effect of man's activities in the area of energy and power? There are three principal influences at work. Firstly, power stations which burn fossil fuels, and indeed all engines which use carbon based fuels, emit carbon dioxide. Secondly, power stations emit water vapour into the atmosphere from their cooling towers; this is also produced directly by combustion of fossil fuels. Thirdly, all energy production and consumption releases heat energy into the environment. ('Dirty' power production can also send large quantities of dust particles into the atmosphere, but this is probably dwarfed by natural dust emitters such as volcanoes.)

Measurements of the atmosphere's carbon dioxide concentration were begun on a systematic basis only during the International Geophysical Year of 1958. It would appear that the concentration of this gas is increasing by about 0.2% per annum, that is 0.7 ppm in the present value of about 320 ppm. Estimates of the amount of fossil fuels consumed each year, and consequently of the amount of carbon dioxide liberated each year, suggest that approximately 50% of the emitted carbon dioxide is retained in the atmosphere, with the remainder being absorbed in the ocean and biosphere reservoirs. On this basis, and with present projections of fossil fuel use, it is possible to show that the atmospheric concentration of carbon dioxide will increase from 320 to 380 ppm by the year 2000. Clearly there must be considerable doubt surrounding this estimate, but it serves to illustrate the full potential of this form of pollution. Indeed, if we use the same model to predict the effect of burning all the known fossil fuel reserves during the next century, the result is a four-fold increase in carbon dioxide concentration. Using extremely crude models of atmospheric behaviour, it is already possible to predict the outcome of such a perturbation. Thus it has been calculated that doubling the concentration of the gas would lead to an increase in surface temperature of 2.4°C. On the basis of the above predictions, this would imply an increase of 0.5°C by A.D. 2000. This is comparable to the 'natural' increase in the first half of this century.

Water vapour emitted from the cooling towers of power stations can become trapped under an inversion layer and thus lead to fog formation. In large conurbations this may be enhanced by the water vapour produced by the combustion of hydrocarbons, including those burnt at the power stations themselves. In some winter conditions the atmosphere will simply be unable to absorb all of this water and foggy conditions will ensue. However, it is unlikely that the water vapour produced in this way will seriously compete with the natural water cycles, at least on a global scale.

Finally, waste heat from power stations, factories, motor vehicles and so on, is a potential influence on climate and, of course, any energy which we use is eventually degraded into this form. It must be remembered, once again, that man's power consumption is some four orders of magnitude less than the solar input, so it is somewhat unrealistic to assume that this thermal pollution is having a significant effect on world climate. However, it is possible that concentrated heat sources, such as large urban conurbations, will influence local climate. Thus it has been estimated that the 4000 square mile area of the Los Angeles basin generates heat energy at a rate only twenty times less than the solar input.

The conclusion drawn from the foregoing is that the present-day amount of power production has a negligible effect on global climate, but that this should not be regarded as an excuse for complacency.

4 Power from Natural Sources

In Chapter 2 we discussed the nature and the extent of the earth's energy resources. Then, in Chapter 3, we dealt with the problems arising from handling energy through the various conversion stages on the way from primary fuel to end use. Now we must turn to filling in the missing section in the mosaic, the production of useful energy from the primary resources. The more conventional techniques associated with the fossil fuels and nuclear power we leave until the following chapters. For the present we shall concentrate our attention on the cinderella power sources of the developed world, the so-called natural power sources.

The magnitude of the sun's energy input and its breakdown into water circulation, wave energy and wind energy have already been described in Chapter 2, where we also discussed the availability and limits of geothermal and tidal power. Some of these power sources have been exploited since earliest times, others are only now being recognised as the soaring price of primary energy forces us to look for 'cheap' and 'renewable' alternatives. The most important one, both in terms of its total magnitude and in terms of the myriad possibilities which are being suggested for its exploitation, is surely the direct use of solar power, and it is with this that we commence our discussion.

Solar Power

It must be constantly borne in mind that most of our energy supply, and all of our existing fuel supply, comes from or originally came from the sun. In this sense all power utilisation is based on solar energy. What is usually meant, however, by the term solar energy, is the direct utilisation of the sun's radiant energy as it reaches the earth. What happens when the sun's radiation reaches the atmosphere? Some of the incident radiation is immediately reflected, and so represents a direct loss of energy to the earth. Some is absorbed into the atmosphere. This eventually turns to heat, setting up convection currents (winds), and supplying the driving force for windmills. Of the radiation that reaches the surface of the earth, some goes to photosynthesis; some is reflected upwards providing for more absorption in the atmosphere; some is absorbed in the ground, heating it up and again causing convection currents in the atmosphere; some goes to heating the oceans causing convection currents and helping to drive oceanic currents; finally, some goes to evaporating water from rivers, lakes and seas, so producing a circulation of water vapour in the atmosphere.

The budget for the incoming flow of solar energy was shown schematically in Fig. 2.1, where an attempt was made to indicate the energy flows and to include the component going towards the production of fossil fuel for the future.

The problem with using solar energy directly is that we have to rethink our attitudes towards energy conversion. We know how to handle our traditional power sources; fundamentally, we burn the fuel, optimising conditions to extract the greatest possible heat, and then convert it into electricity or mechanical movement or whatever, in a secondary device. If we decide to use solar power directly, then we must accept that we cannot tailor the power source, or the amount of heat available from it, by burning more. There is a certain amount available and we can use only that amount. Also, there is a general conviction that, because we can walk around quite comfortably in open sunlight, and that it is only on some parts of the earth's surface that this can become positively uncomfortable, there is not enough energy, or rather it is not hot enough to drive power stations or to heat houses. Nothing could be farther from the truth.

The surface temperature of the sun is 5800°C. This means that the radiation coming from the sun has wavelengths appropriate to this temperature. There is a continuous spectrum of wavelengths present in the radiation, and the maximum intensity occurs at a wavelength that is determined entirely by the temperature of the sun. This arises because the sun behaves very much like a 'black body emitter'. A *black body* is a thermodynamical concept, and indicates a body which absorbs all radiation wavelengths perfectly. If it is at a temperature T and is to remain at that temperature while still absorbing all incoming radiation, it must also be a perfect emitter of radiation and will emit a continuous spectrum appropriate to its own temperature. The different wavelengths will be emitted with different intensities, but a complete spectrum will be present. The wavelength of the maximum intensity radiation is determined only by the temperature of the body. All hot bodies approximate to a black body to a greater or lesser degree, and the consequence is that the colour of a hot body is, to a large extent, independent of the material composing it, and the hotter a body is, the whiter it appears.

The frequency distribution of the radiation from the sun approximates closely to that of a black body of temperature 5800°C, as shown in Fig. 4.1. The complete spectrum of radiation comes from the sun to the earth and arrives at the top of the atmosphere basically unaltered. If it could be collected there by a perfect black body, then the temperature of that body would be 5800°C. However, there is no such thing as a perfect black body. Every material reflects some radiation and in addition convection ensures that further heat is lost from a receiver and so increases the deviation from

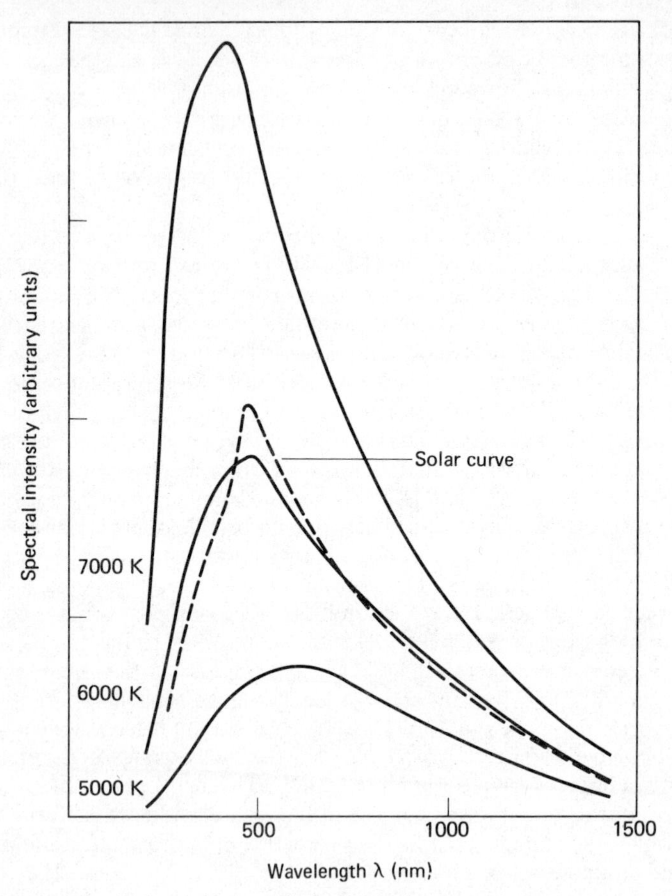

Fig. 4.1 Comparison of solar radiation with that from black bodies at three temperatures.

black body behaviour. Consequently, no real body can be expected to reach this temperature. Nonetheless, large-scale focusing furnaces are practicable, and in particular, the French solar furnace at Odeillo in the Pyrenees has attained temperatures of 3800°C.

A further complication is the effect of the atmosphere which preferentially absorbs certain wavelengths out of the incoming radiation. In particular, the very high energy component in the ultra-violet is almost completely absorbed. This means that the apparent solar temperature as seen at the surface of the earth is slightly lower than that as seen at the top of the atmosphere. The reason why it does not feel so hot is because the energy density is low, and because much of the heat is removed by convection currents. This is readily apparent if one contrasts the sensation of discomfort that is possible when sunbathing in a sheltered spot to that which is felt under the same conditions merely by moving out into the breeze. The additional air movement removes heat from the skin at a much higher rate and so reduces the equilibrium temperature.

There are potentially two ways of using solar power on a large scale. It can be trapped and used to heat some suitable material, or it can be converted directly to electricity using photovoltaic cells. The efficiencies of these photocells are no greater than 20% at the moment (and are normally much less) so that this seems to be a rather inefficient way of using the energy flux. However, it must be remembered that a typical input of solar energy is about 600 watts per square metre, and if the whole population of the world were to consume electricity at such a rate that their standard of living were that of Norway (to pick a smaller and more attainable example than the USA), then only 0.01% of the incident flux would be needed. This illustrates the magnitude of the energy flux incident on the earth. Even so, at the moment solar panels are very expensive. There is an even more serious criticism – the production of silicon cells involves the consumption of so much energy that it would take over forty years continuous operation for this to be recouped. It is clear that a large-scale use of these cells would be an error. There is hope, however, that the energy cost of photovoltaic devices may be reduced and if this is possible then they will have a real role in energy supply.

For heating purposes, it is much more efficient, and also less expensive, to convert the sun's radiation directly to heat using the greenhouse effect. The sunlight with its short wavelength, higher energy component enters the space to be heated through a window. Inside, this radiation is partially reflected and partially absorbed by the contents. These then re-radiate some of the energy at a longer wavelength, which cannot escape through the window again. Thus the losses are cut down, and the temperature of the enclosed space rises rapidly. Everyone knows that car seats become hot when the sun shines. There is one argument based on this that the best way of using solar energy for home heating is to have large windows and thick curtains. The windows should face the sun, the curtains are open during the day so that the heat can be collected efficiently, and they are closed at night to reduce losses. In fact, while this is an efficient system, it has the big disadvantage that the rooms which are used to collect the heat become uncomfortably hot during the daytime.

In practice, it is also possible to use solar energy directly to heat water or air by using simple flat plate collector–heaters. These will work to a certain extent even if they are exposed to the air in the same way that a metal plate becomes hot when left in the sun, but the effect is much enhanced if the collector is covered with a sheet or sheets of glass to reduce convection losses, and also to introduce the greenhouse effect.

The analysis of this system to determine its effectiveness is straightforward. For an incoming power density, P, incident on a black body, the temperature T that the black body will reach is such that the radiation from the body balances the incoming radiation. In practice, this temperature is given by $P = \sigma T^4$ where σ is a constant known as Stefan's constant. In the case of a real body an *absorptivity* α is introduced to allow for the imperfect absorption of radiation, and an *emissivity* ϵ to allow for the re-emitted fraction. Typically α and ϵ are both about 0.9, but it is possible to produce selective coatings which make ϵ about 0.1 for the wavelengths of interest. These coatings are expensive and contaminate easily. If we allow for convection losses by introducing a term which is proportional to temperature, then we can write down an expression for the equilibrium temperature, T, reached by a flat plate collector, as

$$\alpha P = h(T - T_{air}) + \epsilon\sigma T^4$$

If the cover plate is introduced then it will take up a temperature T_c which is above air temperature and less than the temperature of the absorber. The equations are

$$\alpha P = h(T - T_c) + \epsilon\sigma(T^4 - T_c^4)$$

$$\alpha P + P_a = h_c(T_c - T_{air}) + \sigma T_c^4$$

where P_a is the long wavelength component of the incident radiation absorbed by the cover.

This is clearly a complicated set of equations to solve and it will suffice to note that if $h = h_c = 4$ W m^{-2} °C^{-1}, which is reasonable under still air conditions, and the insolation (input solar power density) is 600 W m^{-2}, then the equilibrium temperature of the plate will be 94°C for an ambient temperature of 27°C. There is, therefore, a clear heating advantage compared to about 50°C for the uncovered plate. It is also worth noting that if power is extracted from the absorber, as must be the case if it is being used as a heater, then the temperature of the collector must decrease because there is an additional loss term. Effectively, P is decreased by the amount of power extracted. Fig. 4.2 shows the effect of removing different amounts of power from solar heaters operating at different input powers. It is clear that a compromise must always be reached between the amount of heat removed and the desired temperature. In fact, if the primary consideration is to maximise the amount of heat absorbed, then it should be absorbed at the lowest possible temperature.

One difficulty with solar energy is that it is intermittent. It is unavailable for twelve hours per day on average, and unfortunately in a northern climate, it is available for a shorter period during the winter, when it is needed most. This introduces the necessity for storage in a solar heating system. One possibility is to use solar energy to heat water in one of the flat plate collectors as shown in Fig. 4.3. This water can be taken away to a storage tank and continuously circulated through the solar heater during sunny periods. If necessary it is also circulated through the house, or is used to

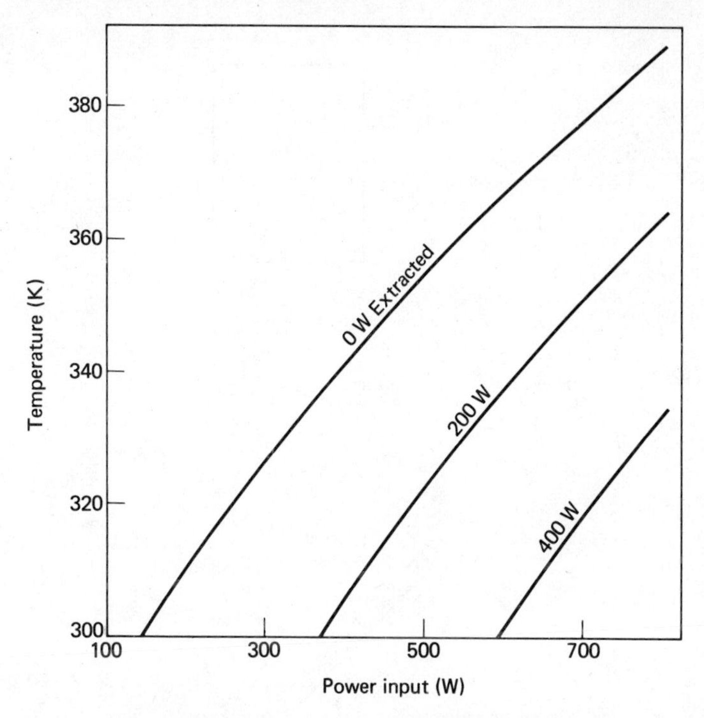

Fig. 4.2 Equilibrium temperature of an irradiated plate as a function of power input for differing extraction rates.

heat a secondary house heating medium, which may be either water or air. At night, or during periods when the solar energy available is insufficient, the flow through the solar heater is stopped, and the stored supply of hot water is used for heating the house. The length of time this store will last depends on two things – the size of the store, and the size of the solar heaters used to heat it. The larger the solar heater, the greater will be the amount of heat collected, and the greater will be the size of storage facility that can be handled. For winter conditions in the British Isles and for a house requiring 10 kW of continuous heating, it would be prudent to include at least five days storage capacity, which represents 52 000 litres (11 500 gallons) of water heated to 50°C. This is a fairly large amount of water and would occupy 52 m^3 (1800 cubic feet) and weigh over 50 tonnes. This basic application, taking the solar heater as a flat plate and capturing what is available, is a very simple way of using solar energy. However, there are more grandiose proposals. Schemes have been suggested in which large areas of desert would be covered with solar panels and used to produce electricity on a large scale. These schemes require the use of focusing techniques to increase the concentration of the incoming solar radiation and hence increase

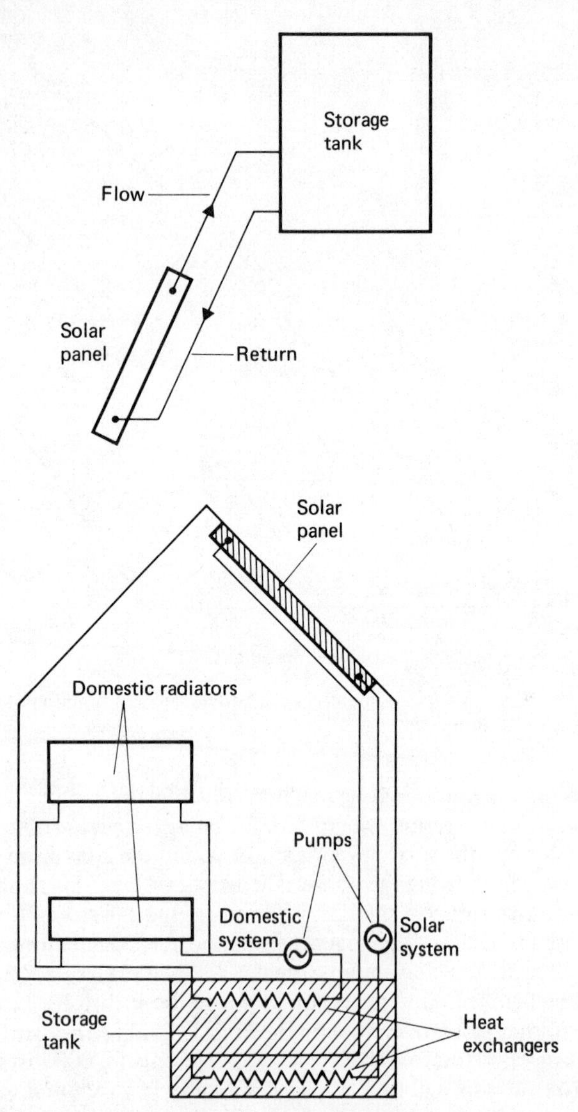

Fig. 4.3 (a) Thermal syphon system: water flows when the top of the solar heater is warmer than the bottom. No pump is necessary, nor any valves, as a cold 'heater' lies dormant.

(b) Pumped system with storage. This system requires a monitor to ensure that the solar system pump is switched on only when the solar panel is delivering heat – otherwise heat will be radiated outwards.

the temperature of the body on which it is incident. This is the same principle as used by school children when burning their neighbour with a lens. Because the term P in our earlier equations is effectively increased by the factor of concentration, the equilibrium temperature must rise correspondingly. Fig. 4.4 shows the relationship between equilibrium temperature

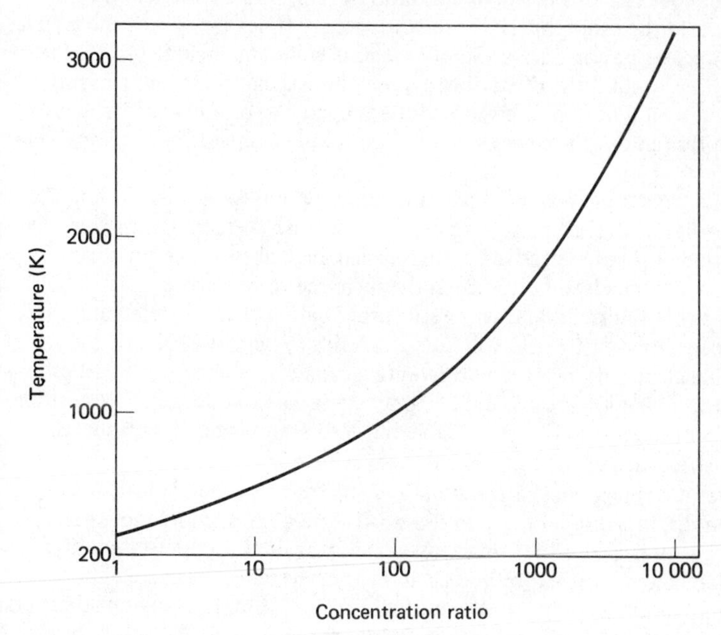

Fig. 4.4 Equilibrium temperature versus concentration ratio for an insolation of 600 W m^{-2}.

and concentration ratio for an insolation of 600 W m^{-2}. One of the electricity producing schemes of this type suggests that enormous tanks would be buried under the desert and the incident solar radiation would be used to raise the temperature of salts contained in the tanks to 1000°C. The hot salts would have heat exchanger pipes buried in them and the coolant in the pipes would be heated and used to produce electricity by conventional means. Perhaps the most esoteric suggestion has been to put a satellite into synchronous orbit. This is a stationary orbit, like those of the communication satellites, in which the position of the satellite above the ground is fixed. This satellite would be a great solar panel some eight kilometres square. Three kilometres away and connected to the panel by an electrical transmission line would be a second satellite. This would consist of a two kilometre square antenna and a converter for changing the electrical input from the solar panels to microwave power. The microwaves would be beamed down to the earth from the antenna and collected in a large receiving station. The solar panel would be capable of intercepting some 8.5×10^7 kW of radiant energy. At an efficiency of about 11% the solar panels would produce approximately 10^7 kW of electrical power. This would be beamed down to earth resulting in about 7×10^6 kW or 7000 MW of power being collected on the surface of the earth. The receiving plant would cover an area of about six times the size of a fossil fuel power station of the same capacity, or about twenty times the size of the equivalent nuclear power station. Safety systems would be included so that if the satellite should drift off station, a controlling signal from the receiving station would not be received and the transmitter on the satellite would abruptly fan out the signal so that it would be dissipated over a very wide area.

The key to the possible success of the proposal for a solar farm in the desert lies in the recent development of selective surface coatings already mentioned. These selective coatings have a high absorbance for solar radiation, and a low emittance in the infra-red region of the spectrum. This means that the sun's light will be absorbed rather than reflected and that, once absorbed, little heat will be lost. In this system, a 1000 MW steam turbine electricity plant would require a capacity of about 300 000 gallons and the collection area would be about 25 square kilometres. The cost of such a project would be formidable but, as its supporters point out, the fuel is free and so the running costs are reduced.

Despite the possible attractions and the romantic grandeur of these proposals, it is undoubtedly in the small-scale domestic and industrial heating applications that solar power can provide the greatest benefit. Careful design and a little forethought in heating systems can make enormous economies in electricity and fossil fuel consumption. It is obvious that solar heating in the Arizona desert is straightforward. The challenge comes in the higher latitudes, in the United Kingdom and Scandinavia, where the solar intensity is lower in the winter months, and where there is extensive winter cloud cover. Even in these difficult conditions there is a saving feature. Heating loads are greatest at times of winter anticyclones. Under these conditions skies are generally clear, and solar intensity at the ground is highest, making solar heating a distinct possibility. Winter depressions are usually accompanied by strong winds and heavy cloud cover, making solar heating impracticable. Fortunately they are also warmer.

Several domestic heating experiments have been made, of which one is the University of Delaware's Solar One house. The roof incorporates two long black panels behind skylights. The heat from these solar panels is conducted away using air, which is used to melt some low melting point salts. At night, air from the interior is circulated through the salt taking up the heat of fusion. In addition, electricity is generated by cadmium sulphide solar cells in the rooftop panels, sufficient to supply all the usual domestic appliances. In the experiment, the house is also connected to the mains electricity supply, but it is estimated that this connection will only be needed for 20% of the time, so showing a considerable saving in fuel generated electricity consumption.

These then are some of the ways in which solar power is being used or can be used to generate energy in a form more readily adjustable to our requirements. The limitations are the efficiency of conversion of the available sunlight to electricity or to heat, and the actual intensity that is available. It has been suggested that the University of Delaware experiment is at high latitudes and that, therefore, the conditions are similar to those prevailing in the United Kingdom. Consequently it is inferred that the lessons of this experiment can be translated directly to the British situation. This is not so. It is not generally appreciated on either side of the Atlantic that Wilmington, Delaware, is at about the same latitude as Madrid in Spain. Putting this another way, the British Isles lie at about the same latitude as Hudson's Bay in Canada. Consequently the problems of Delaware and Northern Europe are somewhat different.

A rather similar experimental house has been set up by the Philips Company at Aachen in Northern Germany. A very sophisticated solar collector is the main source of heat, and a complex system of thermal storage and heat exchangers is used to handle it. When necessary, additional heating is obtained by extracting heat from the ground beneath the house by a mains-powered heat pump (see Chapter 7). Heat loss through the windows is decreased not only by double glazing, but also by a specially developed coating. It is not suggested that all these systems would necessarily be satisfactory in an ordinary family house, even though the minicomputer which runs the systems might make a good substitute for a domestic pet, but as a means of testing developments in the field it is admirable.

The underlying problems in using solar power constructively are these. Can we make photoelectric devices with a sufficiently high conversion efficiency, and with a sufficiently high stability to intense light, to make solar electric converters economically feasible, or is it better to concentrate on thermal collection, perhaps even with a view to generating electricity?

At the moment, typical silicon solar cells can function with a maximum efficiency of about 14% and at a maximum intensity of about ten times maximum solar intensity. Cells based on gallium arsenide/gallium aluminium arsenide are now being manufactured which have a conversion efficiency of up to 24% and can operate at two thousand times maximum solar intensity. It is expected that power outputs of between 20 and 40 W cm^{-2} will be possible. A further advance is that the cell material is more transparent to sunlight than silicon so that a thicker surface layer can be used, which lowers the electrical resistance and increases the maximum power that can be generated. These higher intensity capabilities are important as the incoming sunlight can be focused down from a large collection area by using either lenses or mirrors. This means that much smaller solar cells may be used at the focus of the lens or mirror system. This, in turn, will lead to large economies in design of the collection system for the simple reason that lenses are cheaper than solar panels of the same area.

With the appearance of these cells, the efficiency of conversion of solar energy to electricity is beginning to approach that of conventional power stations. Solar electrical direct conversion systems therefore begin to look attractive. The engineering problems are still immense, and not the least of these is the dissipation of the 75% of the solar energy that is converted to heat instead of electricity.

In the case of heat collection, the greatest problems are in reducing the convection and conduction losses, and in reducing the reflectivity of the absorber across the whole solar spectrum, while ensuring that the re-emission of radiation in the infra-red is also low, that is, in trying to ensure that as much of the incoming energy is absorbed and as little re-emitted as possible. These requirements are conflicting and the techniques so far suggested for solving the problem are complicated. One is to make a surface which is highly absorbing for visible light but highly reflecting for infra-red. This means that the incoming infra-red radiation will be reflected and represents a loss, but equally, the heat generated within the material, underneath the selective coating, will be reflected back into the material. Thus the heat losses are reduced. This type of characteristic absorption and reflection performance can be tailored by using stacked, alternating metal and dielectric sheets. An example of this construction is shown in Fig. 4.5.

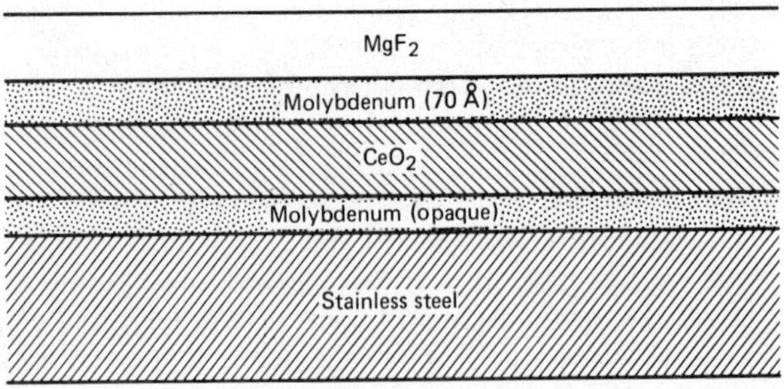

Fig. 4.5 Selective surface for solar heaters. The alternating sheets of metal and dielectric produce interference effects which can be arranged to give a desired absorption profile.

An alternative approach to the problem is to construct a surface, again layered, in which the absorption is achieved in one layer, and the re-emission is inhibited by another layer. These can be created by conventional vapour deposition techniques as used in the semiconductor industry. In one design, shown in Fig. 4.6, a 100 μm thick silicon layer absorbs the visible radiation and allows the infra-red through. A gold layer allows the conductio of energy to the heat exchanger substrate below. This would probably be a

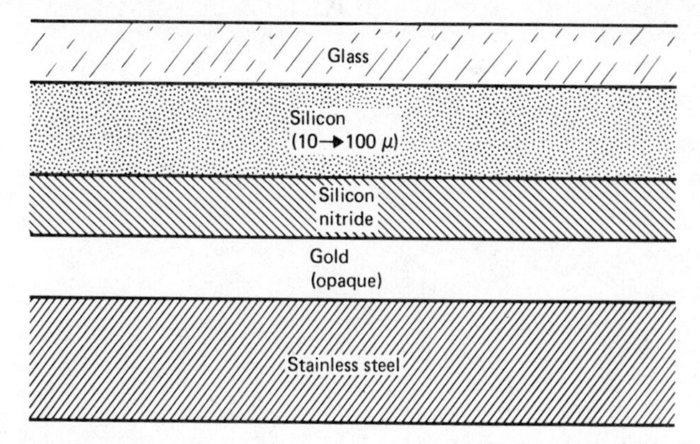

Fig. 4.6 Selective surfaces for solar heaters. The semiconducting silicon layer allows infra-red radiation through but absorbs visible light. The gold layer inhibits re-emission of infra-red and conducts heat to the stainless steel substrate and thence to the coolant. This surface is prepared by conventional vapour deposition techniques.

stainless steel pipe through which the coolant would be circulated. In the proposed power station mentioned earlier the coolant would be liquid sodium which would pass on its heat through other heat exchangers to the salt tanks. The sunlight would be concentrated by a factor of 10 using lens systems and an overall efficiency of 30% is possible at a temperature of about 500°C.

A possible design for the solar energy collector is shown in Fig. 4.7. At the top of the figure is the condensing lens. At the bottom is the energy converter. In order that heat is not removed from the collector by convection, the stainless steel pipe is enclosed in an evacuated glass pipe which is coated on its inside with a reflecting layer except for an entrance slot through which the sunlight enters. The collector is, therefore, like a very long thermos flask. The stainless steel pipe is placed off centre so that light being reflected off the inner surface eventually strikes it. This is a geometrical way of increasing the collector efficiency. Assuming this 30% efficiency such a solar power farm generating one million megawatts would occupy a square of side 120 kilometres.

For domestic applications these selective coatings are also capable of wide application. They would allow the extension of useful solar heating systems into the higher latitudes where efficient collection of the smaller available amounts of incoming energy is of paramount importance. The limitation to their use is their expense. It is only when the industrial demand is such that the commercial price of the coatings is reduced that their application in domestic systems can become widespread.

The direct use of solar power has many immediate attractions. Firstly, it is completely non-polluting. The solar energy is arriving on the earth anyway and it is all destined to become heat. All that we would be doing would be

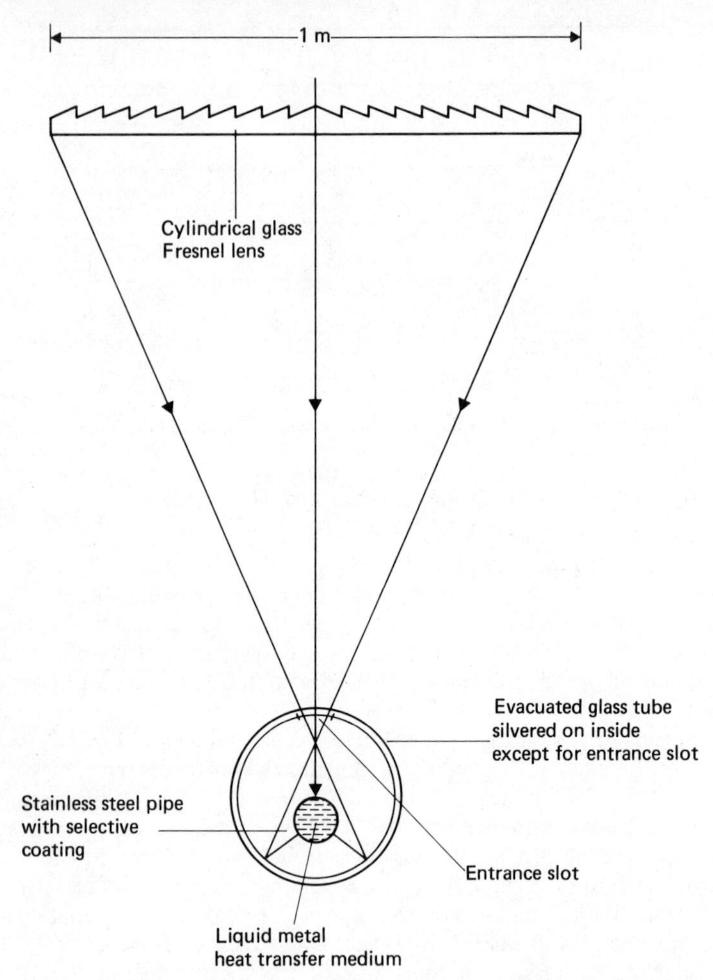

Fig. 4.7 Possible focusing solar collector with selective surface on heated pipe.

to turn it into heat in a way that suited our needs. This heat would subsequently be released and so does not represent a change in the earth's energy balance. The use of solar energy therefore has a twofold benefit: it does not pollute in itself, and it reduces the pollution that would result if the alternative sources were used. Also, the fuel is free and the landscape does not have to be disfigured by the ravages of open cast mining. Against this we must set the requirement that the landscape will be ravaged by the disfigurement of solar farms, but it is *possible* that these would even have advantages in hot arid areas by providing shelter for wildlife and delicate vegetation, so helping to restore plant and animal life in these regions. Solar power farms, therefore, could possibly lead to a rebirth in the deserts. In any event the damage caused by the solar power farms would be less permanent than that due to open cast coal mining. If the farm were to be removed, the land beneath could return to much its previous state, except that it might just have a coating of grass to help keep it together.

The only circumstance under which the use of solar power would lead to a net increase in the energy levels on the earth is the widespread adoption of the suggestion for satellite solar energy collectors with microwave transmission of the energy down to the earth's surface. These collectors would be outside the earth's disc and so would effectively increase the cross-sectional area of the earth by their own area. In real terms, the net increase would be extremely small compared to the actual cross-sectional area and would lead to minimally small increases in energy input.

The most depressing feature of the whole solar energy effort is the minimal investment that has traditionally been made into its research and development. Probably the most important single step in the conservation of the earth's resources has been the decision taken by the oil producing countries to increase the price of oil with such abruptness. The energy consuming societies of Europe and North America have now been forced to spend some money on developing the alternative sources rapidly and economically. For example, in 1974 alone it was estimated that the price rises imposed at the beginning of the year cost the United Kingdom about \$5000M on its import bill. At this sort of cost an investment of, say, \$100M on alternative fuel sources becomes economically attractive. Technically there seem to be no excessive problems. All the necessary technological or scientific breakthroughs appear to have been made, with the exception of the price factor in solar cell fabrication. Economic and commercial feasibility appear to be the only large problems.

The research into solar power implementation would probably be best organised in fairly small groups because of the widely differing energy requirements and problems of regions on a very local scale. Within the United Kingdom alone for example, the problems of the industrial South-East of England, which is the warmest and sunniest part of the country, are distinctly different from those of the North of Scotland with its rural economy and wetter and colder climate.

One further solar device is worth mentioning briefly. A Russian scientist, Alexander Presnyakov, has proposed a solar and magnetic device for the direct conversion of solar energy to mechanical energy. It consists of a wheel whose rim is composed of a ferromagnetic alloy whose Curie temperature lies between 65°C and 100°C. The Curie temperature is the temperature above which the ferromagnetic material loses its strong magnetic properties. Near the top of the wheel is mounted a permanent magnet. As the material on top of the wheel heats up, it loses its magnetic attraction to the magnet as soon as its temperature is higher than the Curie temperature. Therefore, the wheel experiences a net turning force because the rest of the magnetic

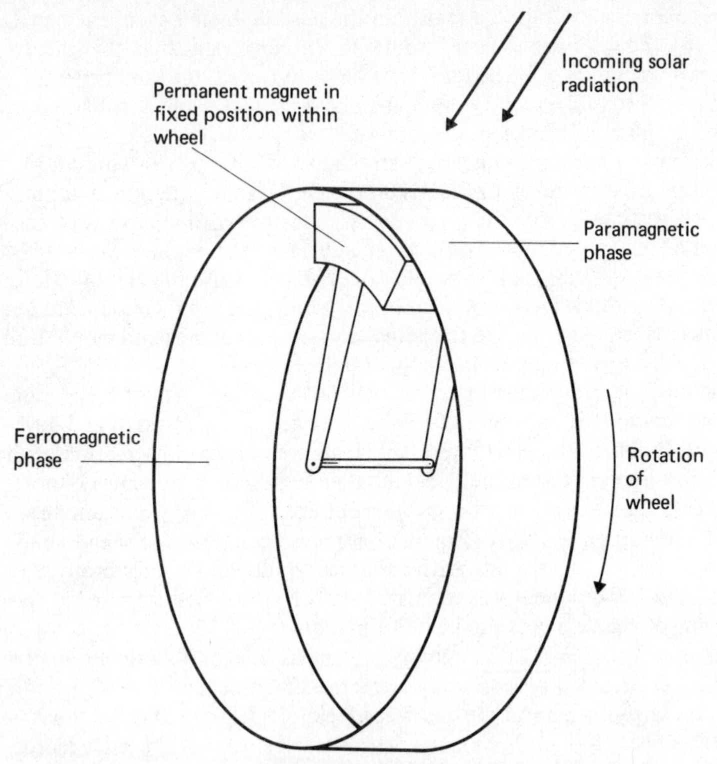

Fig. 4.8 Presnyakov Wheel.

material is still at a temperature lower than the Curie temperature. As a result, the wheel turns, constantly bringing the 'new' ferromagnetic material close to the permanent magnet just inside the rim. This motion will continue as long as the rim is heated sufficiently on one side. The device is shown in sketch form in Fig. 4.8. Presnyakov hopes that one application of his device will be to drive a pump for raising well water.

Water Power

The use of water power by man can be traced back to at least Roman times. It seems that the horizontal water wheel first appeared in the first century B.C. when it was commonly used for cereal grinding. The vertical wheel, which made its appearance in the fourth century A.D., was to become the prime mover behind the industrialisation of Europe after the Middle Ages. Then, during the eighteenth and nineteenth centuries, when its use in Europe was at its peak, it was also to play an important role in North America in textile mills, saw mills and so on. (Fig. 4.9 shows a water power installation

Fig. 4.9 Wellbrook Beetling Mill, Northern Ireland. (Reproduced by permission of the National Trust for Nothern Ireland. Photograph by Sean Walters.)

which is still operational today.) However, the development of hydroelectric power, that is, electricity generated directly from water power, is a recent phenomenon. The first major installation was opened at Niagara Falls, North America, in 1895 with an output of 3750 kW. One of the earliest hydro-electric installations in Europe (1883) was the famous Giants Causeway railway in County Antrim, Northern Ireland.

Today, hydroelectric power is the dominant aspect of water power, and the technology for achieving the conversion to electricity is, in principle, quite simple. Water flowing from high ground towards the sea is dammed at some suitable point so as to produce a large storage lake. This is not actually necessary, as is illustrated in the Niagara Falls example, but it is common practice to dam a slow flowing river and so create an artificially large pressure head or drop for the water. To generate electricity, the sluice gates of the dam are opened and the water forced to flow through a machine which converts the water's kinetic energy into rotational kinetic energy of

Fig. 4.10 The runner of a Pelton wheel. (Reproduced by permission of Boving and Company Ltd.)

a rotor. The rotor is linked to a generator which produces electrical power. The design used depends on the height through which the water falls at the particular installation.

For a large head, say greater than 100 metres, a Pelton wheel is the best choice. This is a flywheel-like structure mounted on a shaft and with a number of cups or buckets fixed around its edge (Fig. 4.10). Each of these buckets has a ridge in the centre which divides the incident jet. The bucket surface is such that the water runs smoothly around it and out at the lower edge without any splashing. The wheel is designed to rotate at such a speed that the buckets are travelling at half the speed of the incident water jet; in this way all the water's energy can be extracted and it falls from the buckets with zero velocity. The capacity of a Pelton wheel may lie anywhere in the range kilowatts to megawatts – the Big Creek installation in California develops 55 MW from a head of 750 metres.

For heads less than 100 metres, or when the quantity of water to be handled is large, a water turbine is used instead of the Pelton wheel. Two types are in current use, the Francis turbine and the Kaplan turbine. Broadly speaking, Francis turbines are used for medium heads, say 100 metres, while the Kaplan variety is used when the head is very small. These latter have movable blades which can be adjusted to maintain high efficiency over a wide range of loads, and some have been developed with runners (the rotating parts of the turbine) over eight metres in diameter.

The main attraction of hydroelectric power stations, apart from their reliance on a natural source, is their high efficiency. Whereas oil- and coal-fired electricity is produced at an efficiency of around 30%, hydroelectricity is produced at 80–90%. And, of course, the basic resource is free. However, hydroelectric installations are extremely expensive to build and this inhibits even the wealthiest governments from investing in schemes with anything but the best prospects. The resource is not only free, it is effectively infinite. There is one small dark cloud on the horizon all the same – silting behind the dam. Upstream from the dam the river or rivers are still flowing freely and eroding the land as they go. Sediment which would normally be carried to the sea may now be deposited in the still waters behind the dam, filling in the lake. A severe build-up of sediment would shorten the useful life of the dam and so make the resource effectively non-renewable.

Wind Power

Just as with water power, wind power has been used by man for many centuries. A few of the windmills which the old Masters included in their canvasses are still to be seen today, though not always in a functioning state. These mills were used to grind wheat or to pump water, the rotation of the large vanes about their horizontal axis being coupled to a grinding wheel or pump. Smaller multivane windmills were common in America and Australia where they were used to raise water from artesian wells. Perhaps the aesthetic peak of wind technology lay in the nineteenth century clipper ships, used to speed tea and other goods from India and Australia to Britain. These ships established speed records which stood until 1976. But, given the way that our society has developed, the ultimate goal must be the conversion of the energy of the wind into electrical energy. Such wind generators were to be seen a decade or more ago but the decreasing price of fossil fuels ensured that their development was short lived. Now the tables have been turned, and once again the wind is being considered seriously as a power source.

We saw in Chapter 2 that the power available from the wind varies as the cube of the wind speed. This rapid variation with speed leads to considerable problems for wind generators. The tremendous increase in power available at higher wind speeds must be balanced against the greatly increased stresses on any piece of machinery (the force exerted on a surface by the wind increases as the square of the wind speed). Thus it may be that the wind generator is unable to take advantage of the very strong winds which can deliver most power. The windmill design, then, must be a compromise between a structure which will be sensitive to very low wind speeds, and therefore an efficient harvester of energy in relatively calm weather, and one which will be capable of operating at the highest wind speeds normally encountered. The compromise inevitably leads to a limit on the upper operating speed. Beyond this the windmill vanes are 'feathered' so as to provide no opposition to the air flow and, hopefully, the structure remains intact. (It is interesting to note that the windmill owner and his wife in former times had to feather the blades manually to ensure the preservation of their livelihood.)

What is the best design for the 'vanes'? Traditional applications of windmills called for low speed, high torque designs. Conversely, in electricity generation applications, high speed is the principal requirement. This leads to the adoption of a radically different design philosophy. A somewhat obvious example is a propellor on a horizontal axis. In theory this should have only one blade for maximum efficiency (60% is the theoretical limit) but dynamic balancing and other problems lead to the two- or three-bladed propellor as the most suitable choice. Unfortunately the efficiency of such an air-screw is rapidly degraded if it is not pointing directly into the wind. Consequently, some sort of tail vane assembly must be used to steer the blades. Moreover, the propellor must be mounted on a tall mast (higher at least than the radius of the blades), and to ensure the strength and durability of such a structure in strong winds is a major engineering problem – to say nothing of the implications for the cost of the device. Nevertheless, such machines have been built. The largest was the 1.25 MW Smith–Putnam generator erected in the United States in 1941. This had a tower 33.5 metres high and a twin-bladed propellor 53.5 metres in diameter. After a few years of trials it entered commercial service as an electricity generator only to be wrecked by strong winds!

Many constructional problems are overcome if the propellor is replaced by a fan with a vertical axis of rotation. Since this is at right angles to the wind direction at all times, steering is unnecessary. Further, the axis of the fan coincides with the axis of the supporting structure which removes the need for complicated gearing and also leads to greater overall strength in high winds. Two simple rotor designs are shown in Fig. 4.11. The first is a simple S shape. The second, the Savonious rotor, allows some of the over-

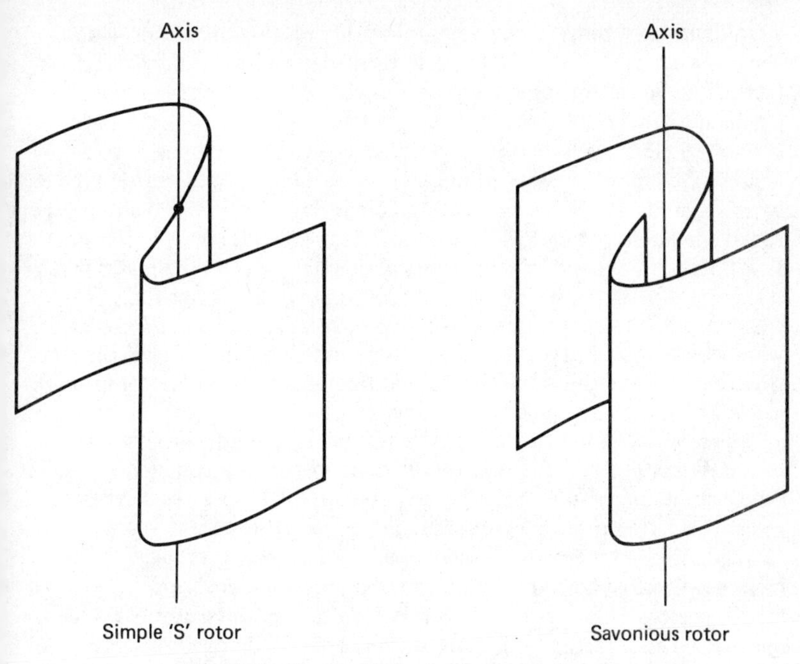

Fig. 4.11 Two vertical-axis wind rotor designs.

spill from the cup facing the wind to be deflected onto, and hence help to drive, the second cup. Neither of these designs can achieve an efficiency to match a well designed propellor type.

Assuming that we have the shaft of the wind generator rotating, how do we then go about producing electrical power? The problem of course is the variability of the power source. This will lead to fluctuations in frequency or voltage or both. There are, however, a number of ways around this problem. To begin with we can use the rotating shaft of the generator to drive an alternator. This will generate a.c. electrical power whose frequency varies with the speed of rotation, and hence with the wind speed. The a.c. can be rectified to d.c. Now we have two possibilities for the next step. One is to use an electronic inverter to give an accurately stabilised a.c. supply of the required frequency. The other is to store the rectified output in batteries or to use it to electrolyse water and store the hydrogen so produced as a fuel. Storage will always be necessary in completely isolated wind power systems to enable the owner to cope during periods of calm weather, but it is expensive, as is the rectification–inversion technique. One simple solution is to use the a.c. power generated by the alternator, whatever its frequency, to drive storage heaters. These will operate no matter what the form of the supply. (In this type of application, it may be easier and more efficient to generate the heat directly in an oleogenerator. This is a tank with an arrangement of stationary and moving paddles which generates heat in a liquid by friction. It is a descendant of the original experimental apparatus for examining the relativity between heat and mechanical energy.)

Nevertheless, we have not suggested a reasonably inexpensive solution to the problem of coupling wind power to the rest of the electricity network. A particularly elegant solution here is the use of the induction generator. If the mains network is large enough (for example, the UK National Grid), then the induction generator will donate power to the mains when the speed of rotation of the shaft of the 'windmill' is greater than the frequency of the mains supply, and will extract power from the network when the converse is true. Naturally, a cut-out can be inserted in the electrical circuits to ensure that the generator is isolated when the wind speed is low and this latter condition prevails.

The economics of wind power depend, of course, on the locality. Because of the v^3 relationship, the capital cost of an installation can be reduced dramatically by the use of slightly more windy sites. Large scale wind generators have not been commonly on the market in recent years but extrapolating from the most recent commercial generators suggests that a high performance machine on a favourable site could cost upwards of $2 000 per kW. This does not compare well with the capital cost of a nuclear installation and, further, due allowance must be made for load factors. A nuclear power plant can achieve a load factor of over 90% (Chapter 6) – that is, it can be in operation for this fraction of its working life. On the other hand, a wind generator depends on an unreliable supply of energy which means that the return on capital is diminished. Moreover, the supply utility must provide adequate storage or standby generation to cope in periods of calm weather (Chapter 3).

Power from the Seas

The exploitation of wave power and tidal power both present enormous technological and engineering problems for the would-be designer of an installation. There is already a tidal power scheme in operation at La Rance estuary, St Malo, France, so this technique is clearly proven. Not so with wave power and the reader should bear this in mind during the following description of a possible system.

Waves, of course, are generated by the action of wind on a water surface. The motion is such that each particle of water moves with constant speed in a circle (in the vertical plane). There is no net translation of the water particles – it is energy and not matter which is transmitted by the wave. To extract power from the waves requires some sort of float device which absorbs waves approaching it but does not transmit a wave behind it. The efficiency with which a given device can do this depends on its shape and on how it is loaded. If the float is fixed, the incident waves will be wholly reflected and no power will be extracted. If the float is free to move, there will be no reflection, but a wave will be transmitted beyond the float and, again, no power will be extracted.

The most promising float to be devised so far is the design by Salter, who proposes a rocking boom with a specific cross section as shown in

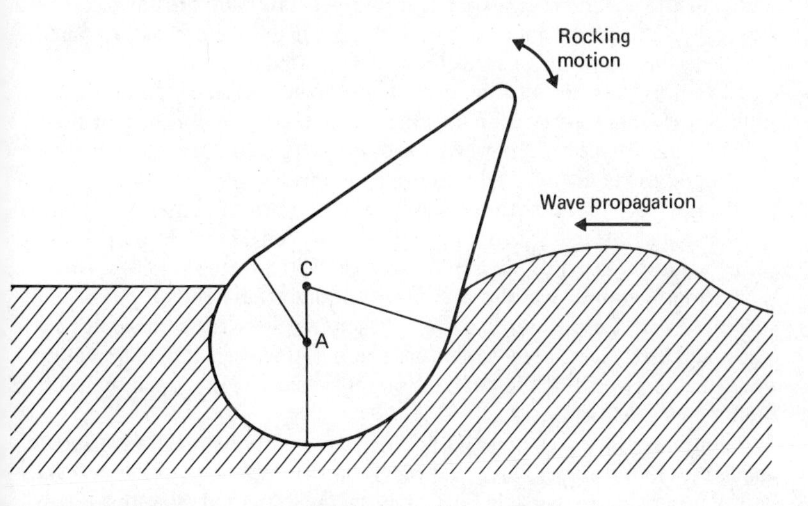

Fig. 4.12 Salter's design for a rocking-boom wave power device. The float rocks about an axis through A. The lower front surface is a cylinder centred at C. (Based on a diagram published in 'CEGB Research'.)

Fig. 4.12. In principle the leading edge moves with the free motion of the water and the trailing edge's circular cross section ensures that no disturbance is transmitted beyond the float. Initial laboratory tests have proved that this float can convert 90% of the wave energy into mechanical work, but these tests were carried out with a regular wave pattern and with the motion of the float electronically tuned to the wave frequency. If more practical conditions are simulated in the laboratory, the efficiency drops to below 50%.

In the harsh environment of the North Atlantic or the North Sea this device would have to cope with large short term fluctuations in wave energy. For a small proportion of the time the waves would be so strong that the device would have to be shut down in some way. Conceivably, this could be done by submerging it! For another small fraction of the time, the wave power would be so small that little or no power could be generated. This would mean that adequate storage or standby power would be needed. This naturally impairs the economics of wave power and it is estimated that the device itself would cost something like \$1600 per kW to install at 1976 prices.

Even if the floating boom can be designed satisfactorily, there is still the problem of getting the power ashore. Because of the slow oscillation of the waves a direct electrical generator on the boom is not an attractive possibility. It has been suggested that the rocking motion could be used to provide pulses of high pressure water by means of a reversing pump. This high pressure water could be used to drive a generator and the power brought ashore by submarine cable. Alternatively, the high pressure water could be pumped ashore and the electrical power generated on land.

If we move further into the realms of fantasy (at least as far as the nineteen seventies are concerned), then it is possible to conceive of floating factories making use of the wave power directly. In the last chapter we mentioned the concept of a hydrogen economy. If this were to become reality the electrolytic hydrogen could be produced offshore. Alternatively, the pumping motion of the floating boom could provide the large quantities of sea water required if uranium is to be extracted from the sea. If this separation technique were to become economic reality then it has been suggested that there would no longer be any need for the somewhat risky fast breeder reactors (see Chapter 6). There is, however, the qualification that the energy cost of such extraction might only be recouped by using the breeder reactor.

To summarise, wave power is a practical possibility on the drawing board and in the laboratory. However, considerable development work is necessary before power in any sizeable quantities will be extracted from the seas around our coasts.

Compared with wave power, where the periodic time of the power source is only a few seconds, tidal power relies on a source which oscillates every 12.4 hours. As we saw in Chapter 2, the mean tidal range can be as high as 10 metres in narrow estuaries, but the range itself varies over a period of approximately 14 days between a maximum 'spring' tide and a minimum 'neap' tide. The spring range may be two or three times the neap range. On top of all this there are smaller seasonal variations over the course of the year, mostly due to the perturbing influence of the sun on the earth–moon system. Now in Chapter 2 it was pointed out that the tidal power which can be extracted varies as the square of the range. Given the substantial variation in this parameter, it is clear that to obtain a steady output of electrical power presents many problems. A steady output is attractive to the public utility whose job it is to supply electricity – an intermittent output merely allows the generating authority to off-load other power stations which must still be available when the tidal power output is low.

As mentioned in Chapter 2, the essence of a tidal power scheme is the construction of an aritificial basin (a dammed estuary) in which sea water can be impounded at high tide. The water is then allowed to escape into the sea at low tide, driving turbines as it does so. By storing all the water until close to low tide, and then allowing it to escape in a short period of time, the turbines are forced to operate under a variable head of water. It can be preferable to circumvent this by adjusting the rate of outflow into the sea in such a way as to maintain a constant head. This increases the generation time and does improve the efficiency of the turbines, but unfortunately less energy is extracted than in the simpler scheme. The overall efficiency of generation can, of course, be increased still further by installing another set of turbines which operate when the basin is filling.

The world's first major tidal-electric installation began to operate in 1966 at La Rance estuary (near St Malo) in France. As Table 2.5 indicates, the mean tidal range is 8.4 metres and the enclosed basin area is 22 km^2. A total of 240 MW of power is generated by 24 separate units giving an annual energy production of 550 000 MWh. Thus the efficiency of the scheme compared to the potential energy production is only 18%. This is acceptably high for tidal power installations. Recently the output has been upgraded to 320 MW, giving a power efficiency of 25%. This figure has been achieved by the introduction of extremely sophisticated turbines incorporating adjustable blades to allow their use during both the filling and emptying operations. In addition they can, if necessary, be used as pumps.

With a single-basin scheme, however, there is always a lack of power production near high or low tide. This can be overcome by the use of a two-basin scheme (shown schematically in Fig. 4.13a) which is admittedly more expensive to build. Sluice gates allow the high level basin to fill in the period between mid and high tide, and are then shut between mid and low tide. The corresponding gates between the low level basin and the sea allow it to empty between mid and low tide, but remain shut between mid and high tide. Electricity generation is by turbines between the high level and low level basins.

As a further sophistication on this scheme, it is possible to replace the complex and costly sluice gates by pumps. There is a considerable waste of energy across the sluice gates – the head loss. The use of pumps leads to a higher level in the upper basin and leads to an overall net gain in energy! This is not something for nothing and does not flout the Laws of Thermodynamics. We are simply pumping across a small head of one or two metres and then generating across a head of seven to eight metres. (The situation is somewhat analogous to the heat pump which is described in Chapter 7.) At spring tides the energy gain can be as high as a factor of 10, while at neap tides it is only about 1.3.

The main problem with the two-basin scheme shown in Fig. 4.13a is environmental. The low level basin has a tide which is much lower than the natural tides along that coastline. Consequently a pleasant estuary bank can be transformed into a stretch of mud flats! To overcome this may require a two-basin scheme of the type shown in Fig. 4.13b. In such a scheme the construction of the barrage is the major item of expenditure so for this particular geometry there is a case for making the basins of unequal area, with the low level basin the smaller. Now if the same mode of operation is used the small basin will fill too rapidly. This could be overcome by following

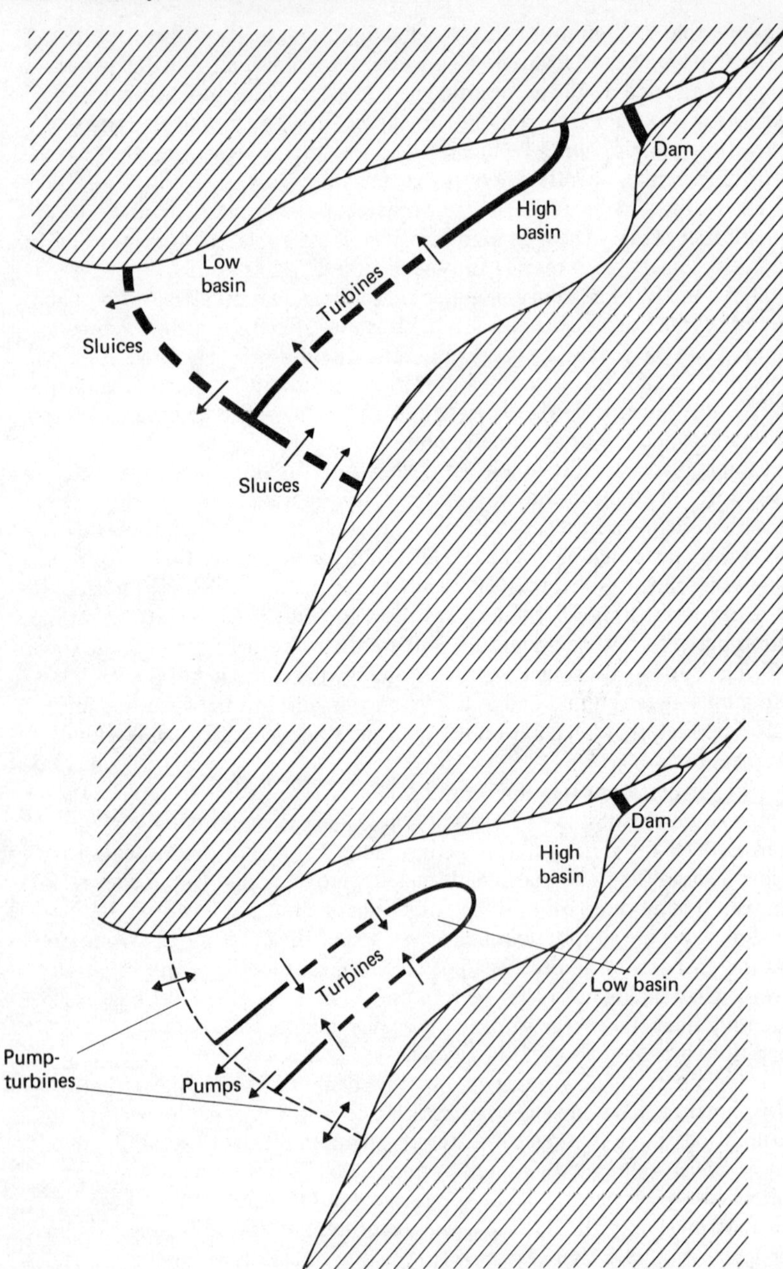

Fig. 4.13 (a) An equal area two-basin tidal power scheme.
(b) An unequal area two-basin tidal power scheme.

the example of La Rance and installing reversible pump-turbines between the high basin and the sea. This permits generation on the ebb tide both between the high basin and the sea and between the high and low basins.

This is the technology for exploiting tidal power, but what of the cost? Many pundits have opined that the costs could be comparable with the installation of equivalent nuclear capacity. However, the construction time for any tidal scheme is likely to be very long indeed and when the ensuing interest charges are taken into account it is unlikely that tidal power can compete economically with nuclear or even fossil fuel power stations. Nevertheless, the resource is free and should become more attractive as fuel costs continue to rise. An additional refinement which might just tip the scales in favour of tidal schemes is that they could be used in pumped storage mode (see Chapter 3). While there might be no extra energy production there would be the usual economic gain through load smoothing by using mains power at off-peak periods to raise the basin level to a higher level than would normally be possible. In fact, the 'heat pump' effect referred to earlier could lead to a net energy gain as the basin height would be artificially enhanced in periods when high tide occurred at strategic times during off peak hours.

Geothermal Power

Heat is continually being transferred from the centre of the earth to the surface. The moderate ground temperatures within a few metres of the surface are in fact maintained by the slow conduction of heat in this way. Unfortunately, the low temperatures involved and the low rate of flow renders large-scale energy conversion unattractive. Heat is also being transported by convection in the hot springs, geysers and molten lava streams which reach the surface in some places. The hot springs can be used directly for heating purposes as is common in Iceland or the steam can be used to generate electrical power. In addition, low temperature water may be available. For example, much of Northern Europe lies over a large mass of ground water heated to a moderate temperature of about 60–100°C and this has already been used directly to heat the Paris studios of RTF, the French television and radio network.

The first geothermal power station was opened in 1904 at the Larderello field in Northern Italy which now has a capacity of 370 MW. The first plant in the United States, with a capacity of 12.5 MW, was opened in 1960 at the Geysers field, California. Typically, the geological structure in the vicinity of a geothermal power source is characterised by a layer of porous rock above a layer of crystalline impervious rock. Stresses in the interior of the earth force molten rock (magma) up into the earth's crust. At depths of about 10 km water in fissures or in porous rocks is heated to as much as 250°C but is still liquid because of the pressures at this depth. Some water may escape through a fissure to boil and flash off into steam. The geothermal energy can be tapped both by using the emitted steam and by drilling down into the porous layers to draw off the superheated water.

The geothermal power installations already in operation throughout the world and their developed capacities are listed in Table 4.1. There are

Table 4.1 Geothermal power stations

	Installed capacity 1969 (MW)	Planned additional capacity (MW)	Earliest installation
Italy			
Larderello	370		1904
Monte Amiata	19		1962
USA			
Geysers, California	82	100	1960
New Zealand			
Wairakei	290		1958
Mexico			
Pathe	3.5		1958
Cerro Prieto	–	75	1971
Japan			
Matsukawa	20	40	1966
Otake	13	47	1967
Goshogate	–	10	
Iceland			
Hveragerdi	(used for direct heating)	17	1960
USSR			
Kamchatka	30.75	7.5	1966

(Reference: M. King Hubbert in Resources and Man, *Publication 1703, Committee on Resources, National Academy of Sciences – National Research Council, W. H. Freeman and Co., San Francisco, 1969. Reproduced by permission of the National Academy of Sciences.)*

essentially three different types of field. Firstly, there are the dry steam fields such as those at Larderello, Geysers and Valle Caldera, Mexico. The steam from such fields can be piped directly to turbines but as its pressure is low (1–15 bars compared with 30–200 bars in a fossil fuel or nuclear plant) large amounts must be used and the effective size of the turbines is limited to about 55 MW. Moreover, as the temperature of the steam is low (250°C compared with 500°C in a fossil fuel plant) the efficiency of power production is poor (this follows from the Carnot formula). Worse still from an engineering viewpoint, the low temperature means that the steam rapidly becomes wet as it passes through the turbines. High velocity water droplets can severely damage turbine blades and special wet-steam turbines must be used. Much more abundant than the dry steam fields are the wet steam fields. Here again the steam can be used in wet-steam turbines to produce electrical power but in addition the hot water can be used for heating or for desalination. Thirdly, there are the low temperature (50–80°C) water fields, most of which appear to be located in Hungary and the USSR. Once again the water can be used directly for heating but in an ingenious plant located at Kamchatka, USSR, the warm water is used to generate electrical power via a heat engine using freon as the working fluid.

We saw in Chapter 2 that the total potential from geysers and hot springs is small compared to the total potential from the whole geothermal reservoir. The problem is to find the technology for exploiting the heat which travels to the surface by conduction. One possible technique is being developed by the Los Alamos Laboratory of the United States Atomic Energy Commission. Water is pumped down a well drilled into the upthrusts of hot rock in the earth's crust. The meeting of the cold water and the hot rock initiates large underground fracture zones in the rocks. By drilling another well close to the first, water can be circulated through the system, heated and used to generate electrical power. A 780 metre deep test bore in New Mexico, using water pumped at a pressure of about 100 atmospheres, succeeded in producing 'hydraulic faulting' of even crystalline rock. If this technique can be developed fully, the usable deposits are increased substantially as the rocks need only be hot and do not have to be permeated with water.

Apart from the technical problems described above, there are a number of environmental problems associated with geothermal power. Small amounts of chemicals such as boron, which are harmful if released, can be brought to the surface, but this can be overcome by injecting the waste material back into the wells. More seriously, land subsidence can occur. In Cerro Prieto, New Mexico, 12 cm of subsidence was reported in 1972, and similar disturbances have been found at Wairakei, New Zealand. In the latter case this is hardly surprising since 70 million tonnes of water are extracted from this field each year – so much in fact that the field has undergone a partial change of character from a wet steam to a dry steam one.

Until the technique of hydraulic faulting becomes practicable, it is clear from Table 4.1 that more geothermal fields will have to be discovered if this method of power generation is to make significant inroads into man's demands for energy. Most exploration proceeds by way of geological studies but infra-red photography of the earth's surface has been introduced recently. Prospects appear to be good. It is believed that large geothermal deposits occur in the Far East, Turkey, along the African Rift Valley and right down the west coast of the American continent from Alaska to Chile. Moreover, Europe has many spas so it would not be surprising if substantial fields existed beneath the whole continent, including the British Isles. We can reasonably expect to see a significant expansion in geothermal power capacity in the near future.

5 Fossil Fuels

The overwhelming dependence of industrialised societies on fossil fuels was amply demonstrated in the introductory chapter, as was the background to the reaction which followed the 'oil crisis' of 1973–4. It is a pity that the situation was allowed to develop in this way, because if governments had paid attention to what some of the world's scientists had been saying for many years, there need never have been any panic. Even now, although there is indeed cause for concern, there is no cause for alarm, because there are still very large reserves of fossil fuels, especially of coal, and it is by no means wrong to continue to exploit them. What is desirable, however, is a change in the nature of the exploitation so that we can obtain greater benefit.

It is worth restating that this was by no means the first crisis in the supply of essential fuel. Prior to the rise of coal, wood was the most important source of fuel. It was essential, not only for domestic heating, but also for all manner of industrial processes, and particularly for the iron industry, where in the form of charcoal it was used both as fuel and as a reducing agent. The great demand for wood led to massive deforestations, with considerable loss of amenity and even, in some cases, severe ecological damage which has persisted to the present day. The shortage of wood compelled the domestic consumer to turn to a substitute, coal, probably with some reluctance, but the problem was far more serious for the iron industry, which found itself unable, for technical reasons, to use coal in its furnaces. It was not until the eighteenth century, when the use of coke in blast furnaces was developed, that the ironworks freed themselves from these strictures.

The foregoing example is not without its lessons for today. Wood can be used for a number of different purposes – as a fuel, as a raw material for industry, and for construction. Today it is essential for house-building, although not for large-scale building and construction work, where it has been replaced by more suitable materials such as steel and concrete. It is essential for the paper industry, but not for iron-making where it has been replaced by the more readily available coke. It would be inconceivable to use it on anything other than a very small scale for fuel, because we rightly regard it as far too valuable to burn. Similarly coal and oil have many different uses, as a source of power for both mobile and static machinery, and as a raw material for the chemical industry. What we need today is a means for assigning priorities to the various uses of the fossil fuels available to us. We have such a means, although it is somewhat imperfect – money. The increasing price of oil, together with political manipulation of its supply, has forced us to reconsider our priorities, and there is still hope that a rational allocation of resources can be reached before the situation becomes really desperate. We may yet decide that coal and oil, like wood, are far too valuable to burn.

Nevertheless, for the next few years we have no alternative but to go on burning them, and it is the purpose of the present chapter to describe the technologies which are used in so doing.

Coal

It is impossible to generalise properly about such a diverse subject as coal mining, but a rather inexact summary will be attempted here.

From the point of view of productivity, open-cast mining is by far the most satisfactory. In this method the overburden (the soil and other materials lying above the coal seam) is stripped off by means of very large earthmoving equipment, and the coal is then removed on an equally large scale. Provided the overburden is fairly thin, this type of mining is very cheap. There are two main problems. Firstly, the whole operation is done where it can be seen, and as it is not visually attractive there are often strong objections to it, especially when it is near to centres of population. Secondly, it is usually necessary to restore the land to something resembling its former state, and while this can be done, it is expensive, and its cost has to be set against the otherwise low cost of mining by this method. Open-cast mining is fairly common in the USA, where the most notable operation of this kind is at the Navajo mine, in New Mexico, which feeds the neighbouring Four Corners power station. In the UK, where the population density is higher and the coal seams are usually deeper, open-cast mining is less attractive.

For deep seams, and for other circumstances where open-cast working is impossible, mining has to be done underground. This is inherently more complicated, partly because of restriction on the use of large-scale machinery, and partly because it is more difficult to get the men in and the coal out. The easiest from this point of view is the drift mine in which the coal seam outcrops at the surface and the mine tunnel follows the seam into the ground at a reasonable angle to the horizontal. In this case both men and coal can travel on a railway. The most difficult is the shaft mine, in which the seam is reached by a vertical shaft, so that both men and coal have to travel by lift.

Underground mining has many associated problems. The roof has to be supported, partly on props and partly on columns of coal which are deliberately left in as roof supports. The mines have to be pumped constantly to prevent flooding, and ventilated, both to provide air for the miners and to prevent the accumulation of methane gas, which could lead to fire. Some coal seams are very thin (less than one metre), and under these conditions the job of the miner is exceptionally unpleasant, and even unhealthy. The sociological problems of mining are beyond the scope of this book, but they may well be the most important limiting factor in the industry, and deserve serious attention.

The chemical composition of coal is complex, and it varies considerably from one type to another. All coals contain a large proportion of carbon, and some volatile matter, and it is possible to classify coal types on the basis of these two quantities, as shown in Table 5.1.

Table 5.1 Coal Classification

Coal type	Anthracite	Semi bituminous	Medium rank bituminous	Low rank bituminous	Lignite
Per cent carbon (by weight)	Over 90	80–90	75–90	70–80	45–65
Per cent volatiles (by weight)	Below 15	15–20	18–26	40–45	Over 45
Burning properties	smokeless very short blue flame	little smoke short luminous flame	smoky long flame		
Energy content (MJ/kg^{-1})	31–33	33–35	31–35	26–31	15–26

In the combustion of coal, most of the carbon and all of the volatiles are oxidised, leaving behind solid ash. In terms of energy this is an efficient process, particularly if the coal is finely divided so that all the carbon is burnt rather than remaining as soot. However, it represents a waste of much valuable raw material which can be used as a chemical feedstock. At present most of the world's chemical industry is based on oil, but with a shortage of oil foreseeable in the near future it is fortunate that coal is available as a substitute.

The chemical uses of coal do not concern us here, but we will proceed to a study of its industrial combustion. As an example we will take the generation of electricity, because it is of great importance and because the techniques are typical of those used in most other large coal-burning plants.

The coal used in electricity generating stations has to be the cheapest available, because the cost of fuel represents about three-quarters of the cost of the electricity sent out from a typical plant, and any reduction in fuel cost represents a major saving. Consequently the coal is often of very poor quality and is in very small pieces, so small in fact that it is quite unusable in an ordinary domestic grate or boiler. It is also important to site the power station fairly close to the coal mines, because the cost of carrying coal is greater than the cost of transmitting the electricity produced. For this reason there is a massive concentration of generating stations along the River Trent in the English Midlands, where both coal and cooling water are readily available, and there is a similar concentration at the Four Corners power stations in the United States, in proximity to the open-cast Navajo mine. Nonetheless, it is not always wise to site the generating plant at the mine itself. Firstly, the coal beneath the plant becomes unmineable because of the risk of subsidence, and secondly, there is a danger that the coal from that mine may become more expensive than expected owing to unforeseeable geological problems. It is politically embarrassing to have to bring in coal by rail from distant mines when the generating plant is sitting on a coalfield, but it does sometimes happen.

Although mechanical conveyors are sometimes used to bring coal from nearby mines, most of that arriving at a generating station does so by rail, usually in trains of about fifty wagons, each holding about 20 tonnes. Special equipment is used to ensure easy and rapid handling of the wagons, from which the coal is taken by conveyor either to storage or straight into the plant. Usually the choice of storage or immediate use is determined simply by the urgency of the station's need, but in some cases the nature of the coal has some importance. Some coals are rather high in sulphur, and may be spontaneously inflammable if stored, and in these circumstances immediate use is essential.

From now on the fate of the coal is as shown in Fig. 5.1. The bunkers feed the coal into a mill where it is crushed to a fine powder. From the mill the coal is blown along in a current of air developed by a fan, until it enters

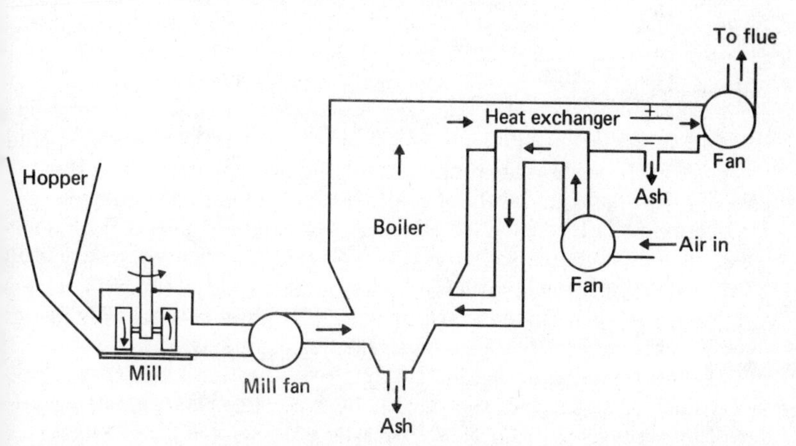

Fig. 5.1 Pulverised fuel plant for large scale combustion of coal. (Modified from McMullan et al. 1976. Reproduced by permission of John Wiley and Sons Ltd.)

the boiler. Here it is mixed with pre-heated air and is burnt, heating the water and raising steam.

Meanwhile the combustion products – mostly carbon dioxide, water vapour and small particles of solid ash – leave the boiler at the top and proceed to a heat exchanger where they give up some of their remaining heat to the incoming air supply. This final step makes good use of almost all the heat left in the gases, and contributes significantly to the very high efficiency (over 90%) which is achieved in a good boiler installation.

Ash particles are removed from the combustion products by electrostatic precipitation. The gases are passed through a mesh between the bars of which there is a strong electric field. The ash is attracted to the mesh and adheres to it. When the mesh is full, the gas stream is diverted to a second precipitator, while the first one is shaken vigorously to remove the ash.

The remaining gases leaving the precipitators are sent up a tall flue and released into the atmosphere. For the most part they are clean, consisting of water vapour and carbon dioxide which are tolerable under normal circumstances, but there is a small amount of sulphur dioxide which arises from sulphur in the coal and cannot be eliminated from the flue gases without great expense. Limitation of sulphur dioxide emissions is thus achieved by choosing low-sulphur coal in the first place. This presents a problem because some high-sulphur coal is very cheap and it is tempting to use as much of it as possible – particularly as sulphur dioxide is invisible and does not usually descend to ground level anywhere near the power station. As a result high emission is unlikely to attract public concern.

Ash from the precipitators and also from the bottom of the boiler is collected in a stream of water and forms a thin slurry. The slurry is pumped out to shallow lakes, known by the exotic name of ash lagoons, in which the ash separates out as a grey sand-like deposit. Until fairly recently the disposal of this ash presented a serious problem, and it was usually left in the lagoons. As a consequence these steadily accumulated around the larger power stations. There are now several uses to which the ash is put, thus turning an embarrassing problem into a saleable commodity. The most notable of these are in land reclamation and as lightweight concrete aggregate. In land reclamation the lagoons are built on low-lying riverside land where the deposited ash raises the level out of the reach of floods. After suitable fertilisation the ash is ready for planting a few carefully chosen crops, and when the humus content of the new land has reached a satisfactory level, it becomes suitable for more-or-less normal agriculture. In lightweight concrete aggregate, ash replaces the more usual sand. The product is not only lighter than normal concrete, but has a lower thermal conductivity, making it suitable for insulation of buildings – thus saving heating fuel!

On a much smaller scale than electricity generating plant, but of considerable economic importance, are domestic and small industrial heating plants. Although most of this market has been invaded by oil, especially in industry, there is still appreciable room for coal. The pleasant aspects of domestic heating by coal have been emphasised in the United Kingdom by the advertising industry ('Welcome home to a Living Fire.'), but there is no doubt that the disadvantages of solid fuel have been reduced by modern coal burning appliances.

The simple open fire is inefficient because only its radiant heat output is used to heat the room, while the large amount of convective heat produced is sent up the chimney. As a result, the efficiency of an open fire is no more than 10%, and is often less. In addition the inefficient combustion conditions lead to smoke emission unless smokeless fuels are used.

The efficiency of heat transfer can be improved by taking the flue gases through a heat exchanger where the heat is transferred to the air of the room to be warmed, and this not only raises the efficiency to perhaps 50%, but also helps to spread the heat around the room by convection. The design of a typical domestic stove of this type is shown in Fig. 5.2. The psychological objection to this type of stove is that 'you can't see the fire', but this is overcome to some extent by means of glass doors. However, it must be admitted that the result is less pleasing than an open fire.

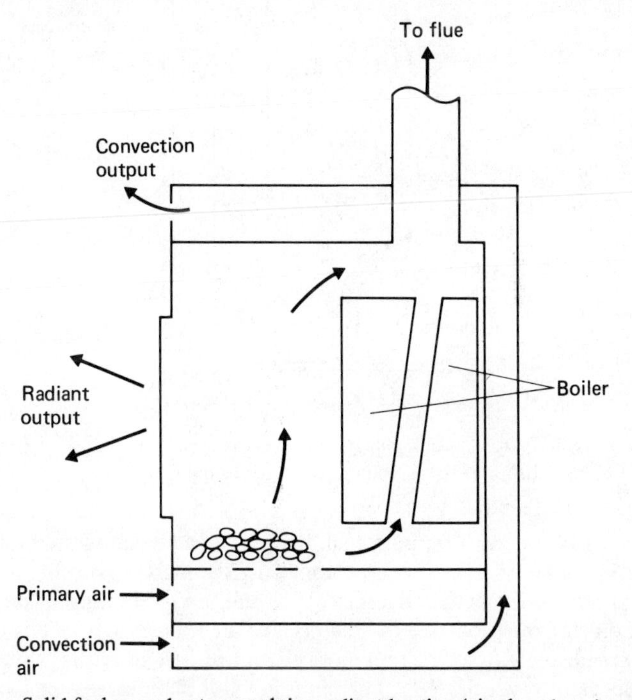

Fig. 5.2 Solid-fuel room heater supplying radiant heating (via glass doors), convection and hot water. (Modified from K. V. Bradbury, *Solid Fuel in the Home*, Women's Solid Fuel Council 1973. Reproduced by permission of the Women's Solid Fuel Council.)

A more satisfactory solution to the problem is to abandon the idea of seeing the fire at all and to confine it to a boiler in a central heating installation, when efficiencies of 70% can be achieved, and some measure of automatic stoking and of thermostatic control are possible. A typical solid fuel central heating plant is shown in Fig. 5.3.

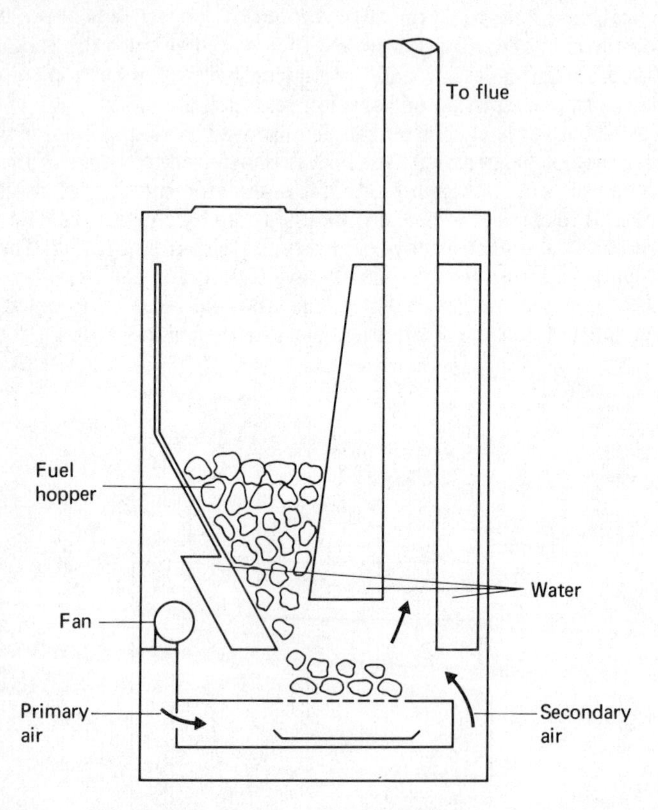

Fig. 5.3 Solid-fuel central heating boiler. (Modified from Bradbury, 1973. Reproduced by permission of the Women's Solid Fuel Council.)

As an alternative to burning coal directly, it can be converted into other fuels which are in some way more convenient either to transport or to use. Examples include the manufacture of gas and the production of oil from coal. Some of the processes to be described are not actually in current use; some are obsolete or at least temporarily out of use, and others, though feasible, are not yet developed beyond the pilot plant stage.

The manufacture of gas from coal was widespread in the United Kingdom from the nineteenth to the middle of the twentieth century. It has been superseded, first by gasification of oil, and then by natural gas. However, these substitutes will not last indefinitely, and since coal is relatively plentiful it is possible that interest in coal gasification will revive. America, blessed with great reserves of natural gas, has not manufactured significant quantities of gas from coal for many decades except as a by-product of chemical processes and of the production of metallurgical coke, but now that American supplies of natural gas are becoming inadequate, it is possible that even the USA will have to come to terms with that landmark of all British towns, the gasworks.

The traditional process in the United Kingdom for manufacture of coal gas involves heating coal in a closed retort. The gas produced is mainly hydrogen, methane, nitrogen and carbon monoxide, and the main by-product is coke, which is a convenient smokeless fuel. Coal gas is a fairly satisfactory fuel. It burns with a high flame speed, and because of its relatively low calorific value it requires a fairly small admixture of air, so it can be burnt in simple burners without difficulty. The disadvantages are high production cost (in comparison with natural gas), high transport cost because of the low calorific value, and risk of poisoning because of the carbon monoxide content.

Another method of coal gasification involves blowing air over hot coke or coal. The carbon partially burns in the oxygen, yielding carbon monoxide, while the nitrogen remains unaffected. The reaction product, producer-gas, is of low calorific value because of the presence of a large proportion of incombustible nitrogen. Its main use is as an industrial fuel.

If steam is blown over hot coke, the following reaction occurs:

$$C + H_2O \rightarrow CO + H_2$$

The product, water-gas, is a mixture of hydrogen and carbon monoxide, and as both these gases are inflammable, the calorific value is higher than that of producer-gas. Unfortunately the reaction is endothermic, so the coke bed must be heated continuously. Alternatively the producer-gas process (which is exothermic) can be cycled sequentially with the water-gas process, alternately warming and cooling the coke.

A variation of this procedure, known as the Lurgi process, combines the producer-gas and water-gas processes by using a stream of air mixed with steam, thus avoiding the need for external heating of the reaction vessel or for a cyclic process. The gas is suitable for town gas supplies, for driving engines and gas turbines, and for chemical synthesis, and because it can incorporate a useful degree of sulphur removal, it can reduce the atmospheric pollution which usually results from coal-burning plant. Recently a modification has been proposed which uses a fluidised bed, in which the coal particles are in aerodynamic suspension in a fast stream of air and steam.

Two methods have been devised for producing petroleum-type products from coal. The *Bergius process* invented in 1913 and in commercial use in Germany from 1927, is a method for hydrogenating coal to produce liquid hydrocarbons. Pulverised coal and hydrogen gas are made to react in the presence of catalysts (tungsten and molybdenum sulphides) at high temperature (~400°C) and pressure (~200–300 bar), and the products can be separated using the usual techniques of oil refining. The process is considered obsolete, but it is of historical interest because it supplied much of the demand for petroleum in Germany during the Second World War. The *Fischer–Tropsch process* is a modification of the water-gas process in which CO and H_2 are made to react to form hydrocarbons. The water-gas is passed at a temperature of 200–300°C and a pressure of 20–30 bars over an iron catalyst. It is a satisfactory means of producing not only fuel but also chemical feedstocks, and in view of the threatened shortage of oil, it may well prove very useful in the future.

Peat

As we saw in Chapter 2, peat is available in quite large quantities, although only a few countries have really worthwhile amounts. The most vigorous exploitation has been carried out in the USSR and in Ireland.

There is considerable harvesting of peat on an 'amateur' basis by individual families for their own use, but mechanical harvesting accounts for the bulk of world production. For example, the Irish Peat Board harvest by machine on a large scale (see Fig. 5.4) and sell the product, often in a compressed briquetted form to individual households in much the same way as coal. In Ireland it is also used on a large scale for the generation of electricity – on such a scale, in fact, that in 1972–3 peat-fired stations provided 29% of the Irish Republic's electricity.

Mechanical harvesting wins the peat either as sods (that is, pieces of a fairly uniform size, approximately 300 x 200 x 100 mm) or as milled peat, which is light and powdery. Sod peat is burnt on a travelling grate which carries unburnt peat into the combustion zone and takes the ash out at the other side. Milled peat is remilled (and partially dried) before burning, in a manner similar to the production of pulverised coal, and thereafter is handled similarly to pulverised coal. In either case the boiler, turbine and alternator are very much like the corresponding equipment in a coal station.

Petroleum

Oil and gas are generally found in association with each other. Since they are both fluids, extraction is less of a problem than with coal, but set against this, the more stringent geological requirements make deposits more localised and therefore more difficult to locate. It is not possible at present to tell from measurements at the earth's surface whether or not oil is actually present underground. It is possible to locate suitable sedimentary rocks and suitable traps in which oil might be expected to have accumulated, and to

Fig. 5.4 Peat winning machine. (Reproduced by permission of Bord na Móna.)

assess the likelihood of oil actually being there, but the actual finding of oil must still be done by drilling. Deep drilling is very expensive, and it is very important, therefore, that the methods used to locate possible oil reserves are as good as possible.

The first stage of prospecting involves the production of a detailed geological map of the area. This is usually done by aerial photography followed by geological examination of the surface rocks, although clearly this is not possible if the area is under water. The next stage, which is necessarily the first stage for underwater areas, is geophysical prospecting, in which the magnetic, gravitational and seismic properties of the area are measured. Seismic surveying is perhaps the most successful. In this technique, a small explosion generates sound waves which travel through the rocks and are reflected or refracted at discontinuities in the geological structures. The reflected and refracted waves are received at a series of geophones and they are recorded continuously over a period of time (usually a few seconds). Interpretation of the results is complicated, but the technique provides more detailed information than other methods. The technique was originally developed for land but it has been modified very successfully for marine use.

When all the surveying has been done, and likely drilling positions established, an exploratory well or wildcat is drilled. The main purpose of this well is not so much to find oil, although of course one is not displeased if one does so, but rather to investigate the nature of the rocks and to confirm or modify the results of the surveys. It is important, therefore, to take samples of the rocks at frequent intervals, both by analysing the pieces of rock brought up by the lubricant (see later), and by bringing up cores – cylindrical samples of the rock through which the drill is passing. In addition a continuous record is kept of the electrical properties of the rock.

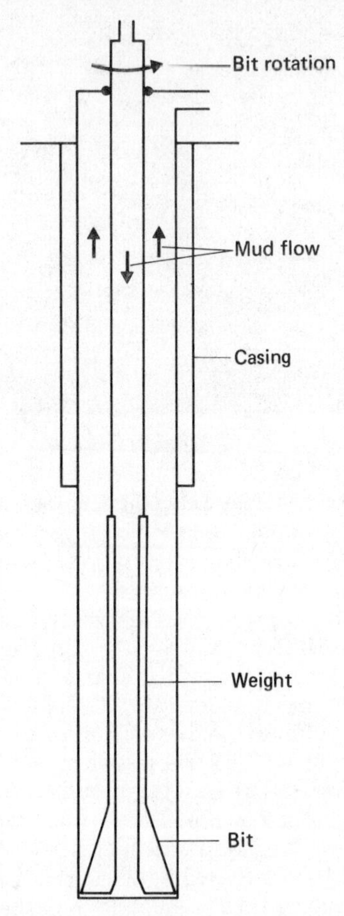

Fig. 5.5 Schematic diagram of oil drilling.

The drilling process is illustrated schematically in Fig. 5.5. The bit, which is tipped with diamond, is rotated continuously, and the shaft is lowered so as to keep a steady pressure on the bit. The shaft is lengthened at intervals. During the drilling, the bit is cooled and lubricated by a stream of specially formulated 'mud' which also serves to bring the cut pieces of rock to the surface for analysis.

When the hole has reached a predetermined depth, the drill is withdrawn and a tubular casing is lowered into the hole and fixed with cement. A valve known as a blowout preventer is attached at the top of the well in case the drill should hit oil or gas unexpectedly, and the drilling then continues with a smaller diameter bit.

When the drilling bit enters an oil or gas reservoir, there may be an increase in pressure. This is corrected temporarily with the blowout preventer while the density of the mud is increased to compensate for the pressure of the oil in the reservoir. Drilling can then continue while the geology is tested. If the well proves worthwhile, it is completed by cementing in the last section of casing.

A series of exploratory wells is drilled to establish the extent and the characteristics of the new oilfield, and when the exploration is completed, production wells can be drilled in much the same way.

The recent oil and gas wells in the North Sea have been drilled by techniques which are similar to the above, but the platforms from which the drilling is done call for some attention. There are two important types. *Self-contained platforms* are set up permanently on piles driven into the sea bed. They are relatively inexpensive, but they cannot be used in water deeper than about 30 m. *Mobile units* are specialised 'boats' which can be towed into the required position and can work in great depths. Most types are designed to put down legs (or even the hull itself) on to the sea bed, but if the depth of water exceeds 100 m this is impossible, and the drilling has to be done from a floating platform anchored in position. At present the limit for this type of platform is about 200 m. Such a platform is shown in Fig. 5.6.

Oil is extremely unevenly distributed about the world, and so when the crude oil emerges from the well it has to be transported over long distances, usually by sea. The specialised oil tanker is by no means new – the first was launched in 1886, and had a capacity of 3000 tonnes. There have been spectacular increases in size in recent years, and vessels of 200 000 tonnes are now by no means uncommon, but it is likely that constraints such as the shapes and depths of waterways and harbours will prevent much further growth in the size of vessels.

Large tankers provide an extremely economical method of transport, partly because the long thin shape of the hull makes for very high efficiency, and partly because the number of crew is often no more than on a smaller vessel. On the other hand they are less useful for other purposes than are most vessels, and the recent contraction in the oil business has led to many vessels being laid up, or at least operating at very much lower productivity than usual.

Pipelines are an alternative to sea transport in many cases. They are rather more expensive to run, and are totally inflexible in that they cannot

Fig. 5.6 Oil drilling rig 'Sea Quest'. (Reproduced by permission of Harland and Wolff Ltd.)

be re-routed in response to changing demand. Nevertheless, there are circumstances where they are desirable or even essential.

Before use, crude oil is refined into a wide variety of products by what are essentially chemical engineering methods. Crude oil is a complex mixture of variable composition, but in general terms it is possible to describe it as a mixture of hydrocarbons with on average about 85–90% carbon, 10–14% hydrogen, a variable amount of sulphur from 0.2–3% or higher, small traces of vanadium and nickel and very small traces of other elements. The hydrocarbons can be classified according to the arrangement of the chain of carbon atoms (see Fig. 5.7). The majority are paraffins, in which the carbon chain is joined by single bonds. The paraffins may be either straight-chain or branched-chain types, and the different types of chain make the compound suitable for different purposes. A different arrangement of the carbon chain is found in the cycloparaffins, which are present to a lesser but still appreciable extent. The least plentiful of the hydrocarbon types is the aromatic group in which the number of hydrogen atoms is rather less than in the corresponding cycloparaffins. There are also small quantities of non-hydrocarbon compounds, especially sulphur compounds. It must be emphasised that the composition of crude oil varies greatly from one producing area to another, but a fairly typical analysis is shown in Table 5.2.

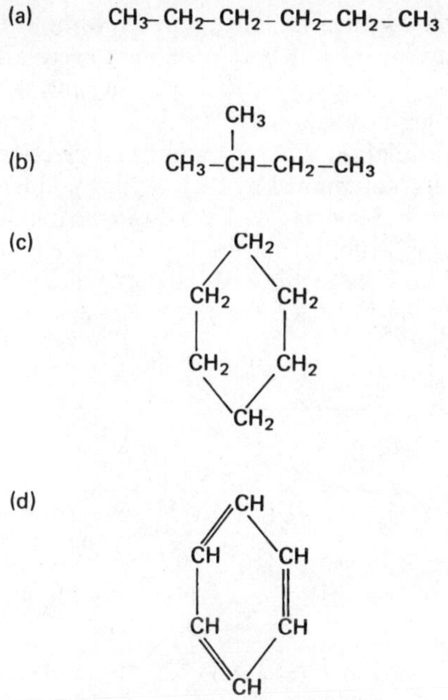

(a) $CH_3-CH_2-CH_2-CH_2-CH_2-CH_3$

(b)

CH_3
|
$CH_3-CH-CH_2-CH_3$

(c)

CH_2
CH_2 CH_2
CH_2 CH_2
CH_2

(d)

CH
CH CH
CH CH
CH

Fig. 5.7 Some typical hydrocarbons in petroleum. (a) Normal or straight-chain paraffin (normal hexane); (b) iso- or branched-chain paraffin (iso-pentane); (c) cycloparaffin (cyclohexane); (d) aromatic (benzene). Note that the bonds in the benzene ring are in fact equivalent, rather than alternately double and single.

The first stage of refining is distillation, in which the temperature of the crude oil is raised steadily so as to drive off the fractions in order of increasing boiling point. This process separates different products to some extent, but is only the beginning. The fractions resulting from straight distilling

Table 5.2 A typical crude oil composition

Constituents	Weight per cent
normal paraffin hydrocarbons	23.3
iso-paraffin hydrocarbons	12.8
cycloparaffins	41.0
aromatics	6.4
aromatic cycloparaffins	8.1
resins asphaltenes	8.4
total	100.0

Modified from M. A. Bestougeff, in Fundamental Aspects of Petroleum Geochemistry, *Nagy and Colombo (editors), Elsevier 1967. Reproduced by permission of Elsevier Publishing Company.*

are often unsuitable for their intended purpose without further treatment, and the proportions in which they occur do not necessarily coincide with the proportions required by the market, and for both these reasons further chemical processing is needed.

The object of distillation is to separate the crude oil into several fractions, known as cuts, determined by their boiling points. There is no hard and fast rule for the boiling ranges of the different fractions, for much depends on the composition of the crude and on the requirements of the processes which follow distillation, but an approximate split might be as given in Table 5.3.

Table 5.3 A typical distillation split

Boiling range °C	Product	Uses
–10 to 0	Propane and butane	Domestic cooking, 'clean' industrial purposes.
25 to 80	Gasoline (Petrol)	Vehicles.
80 to 180	Naphtha	Cracking into petrochemical feedstock, gasoline, town gas.
150 to 250	Kerosene (paraffin)	Heating fuel, aviation.
250 to 350	Gas oil (diesel)	Diesel engines, heating fuel.
Over 350	Residual	Large engines, power stations, large heating plants, lubricating oil, bitumen.

The distillation is carried out in towers whose design is illustrated in Fig. 5.8. The object of this design is to maximise the separation of the different fractions. The tower is heated by a stream of superheated steam, which creates a temperature gradient. Crude oil is heated and pumped into the tower. The more volatile components evaporate and travel up the tower until they reach a region at a temperature below their boiling point, while the less volatile components travel as liquid down the tower until they reach a region at a temperature above their boiling point. Liquid fractions are taken off at different levels corresponding to different boiling points, and the most volatile components continue up the tower and out at the top. The liquid fractions can be subjected to re-distillation in smaller columns to improve the separation still further.

To increase the yields of the commercially important lighter fractions (especially gasoline), and also to improve the octane number of the product, a portion of the heavier fractions is subjected to a catalytic process in which the oil and steam are fed into a reaction vessel (Fig. 5.9) in which there is a fluidised bed of fine particles of fuller's earth. This promotes a very rapid and effective reaction in which the large molecules of the feed

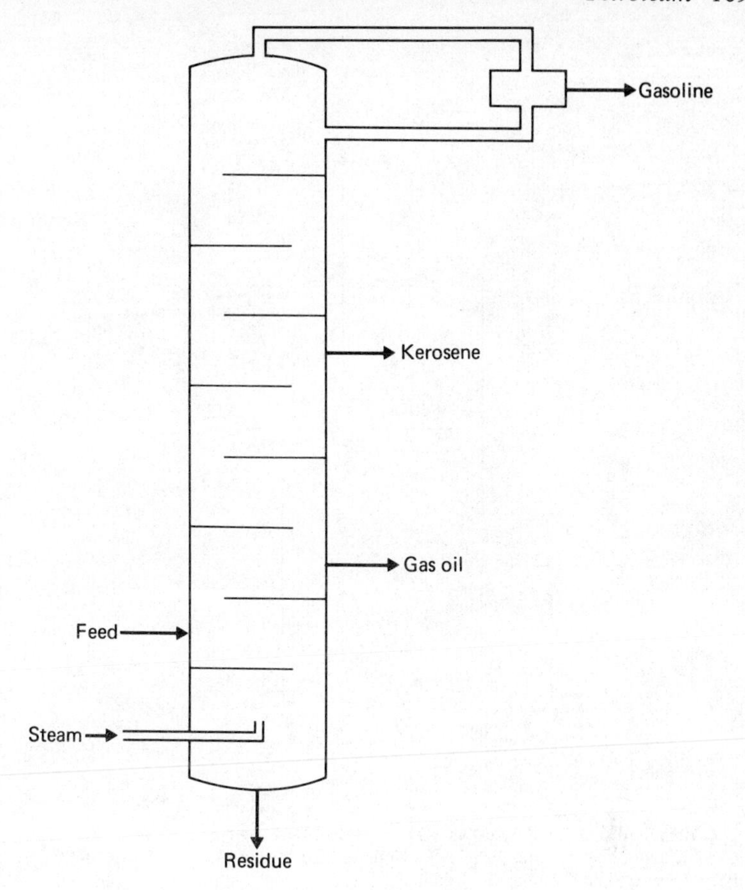

Fig. 5.8 Distillation column.

are broken down into smaller molecules. Some carbon is deposited on the catalyst, and this leads to a decrease in its effectiveness, so it is continuously removed and regenerated in a second fluidised bed (Fig. 5.9) in which the carbon is burnt in a stream of air. The resulting carbon dioxide is released to the atmosphere.

The presence of sulphur in oil fuels is undesirable because it can corrode engines, and because it produces sulphur dioxide which is a rather noxious atmospheric pollutant. Most of the sulphur compounds in the lightest fractions are mercaptans, which are soluble in sodium hydroxide and can therefore be removed by washing with caustic soda solution. The sulphur compounds in the heavier distillates, especially the gas–oil fraction, are chemically different from those in the light fractions and have to be treated

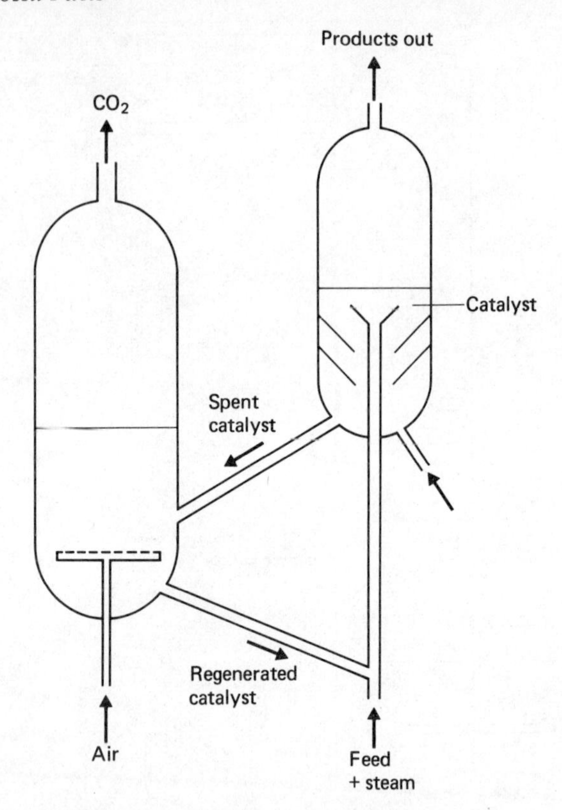

Fig. 5.9 Fluidised-bed catalytic cracker. (Modified from M. H. Lowson (editor), *Our Industry–Petroleum*, BP, 1970. Reproduced by permission of the British Petroleum Co. Ltd.)

by a different process. The sulphur-bearing oil is made to react with hydrogen at high temperature and pressure in the presence of a catalyst of cobalt and molybdenum oxides dispersed on a base of alumina. The sulphur combines with the hydrogen to produce H_2S.

The processes are not suitable for the residual fuel oils, which is particularly unfortunate because this fraction usually contains a rather large percentage of sulphur. Residual oil is usually burnt in large industrial furnaces and power stations, which can use tall chimneys to dispose of the resulting sulphur dioxide, but there has been increasing concern about the effect of this discharge on the environment, and there has been increasing pressure on the oil refiners to reduce the sulphur content of their product. One 'solution' is to use crudes which are low in sulphur, such as those from Nigeria and the North Sea, but clearly this does not solve the problem.

Let us now turn from refining to utilisation.

Petroleum gases

The lightest fractions of petroleum are gases at ordinary temperatures. The one with the lowest boiling point is methane, CH_4, which boils at −162°C. Methane is the principal constituent of natural gas. Propane and butane, which are the familiar bottled gases, are known to the industry as liquefied petroleum gas (LPG). Although both propane (C_3H_8) and butane (C_4H_{10}) are gases at ordinary pressures and temperatures, they can be liquefied by compression to moderate pressures and can be stored safely and conveniently in steel cylinders, or even, in the case of butane, in specially designed disposable containers which are little more than 'tin cans'.

The most important characteristic of petroleum gases is that they burn with a hot and very clean flame, making them ideal for all applications which demand cleanliness. These include industrial processes such as the firing of ceramics in which the burning gases are in contact with the product and also domestic cooking where a smoky or smelly flame would be unacceptable to the consumer.

Gasoline (petrol)

Gasoline is liquid at ordinary temperatures, but evaporates easily in air to form a highly inflammable mixture. It usually consists mainly of hexane (C_6H_{14}), heptane (C_7H_{16}) and octane (C_8H_{18}). These hydrocarbons can occur in different isomers, having the same overall proportion of C and H but a different structural arrangement. The most critical part of the operation of a petrol engine is the ignition and burning of the mixture in the cylinder. The mixture is highly compressed, and there is a tendency for it to ignite spontaneously and explode rather than burn smoothly. This is called knocking or pinking. An explosion is both damaging and inefficient, and to avoid it, the nature of the isomers in the fuel has to be controlled. In general the straight-chain isomers, and especially normal heptane, are poor in this respect, while the aromatics, and the branched-chain isomers, especially iso-octane, are good.

The octane rating of motor fuels is generally enhanced by the addition of an anti-knock agent such as tetra-ethyl lead, but with the growing public concern about lead pollution from car exhaust, the extent of this practice has been reduced. Consequently it has become much more important to control the octane rating by methods such as catalytic reforming, which unfortunately make the fuel more expensive.

The octane rating of a fuel is of much greater importance in engines with a high compression ratio, and it might appear that the solution to the problem of lead pollution is merely to use engines with a low compression ratio, which can thus use fuel without the addition of tetra-ethyl lead. However, this suggestion raises two problems. The first is that leaded fuel deposits a thin layer of lead on the cylinders and valve gear of the engine, which is an important part of the lubrication system. This can be rendered unnecessary by a change in engine design. The second and more serious problem is that the efficiency of a petrol engine increases with compression ratio, so that a car with a high compression ratio has a lower fuel consumption than a similar car driven in a similar manner but equipped with a low compression engine. In the interest of fuel economy, therefore, it is essential to use high compression ratios, and if the addition of lead compounds is prohibited, the only alternative is to use techniques such as catalytic reforming.

Kerosene (paraffin)

This is obtained from straight distillation in the boiling range 150–250°C. Originally it was primarily a fuel for lamps, but its main use nowadays is for heating, and especially for domestic central heating systems. The quality rating of a kerosene is therefore determined by its burning qualities. To burn a liquid fuel it is first necessary to vaporise it. Unlike petrol, kerosene shows no tendency to evaporate at room temperature, and it must be encouraged to vaporise by heating just before combustion. In a simple wick-fed burner the fuel takes in heat from the flame as it travels up the wick, and it evaporates at the surface of the wick in a zone which can be seen clearly in a candle flame. Various other methods of vaporising have been developed, some of which involve mechanical agitation. An example of this is the wall-flame burner, which is the standard type for domestic central heating boilers. Its construction (Fig. 5.10) is more complicated than the simple wick-fed burner, and it requires an electricity supply for ignition and to drive the spray and air vanes, but it is amenable to automatic control and it is easily capable of the necessary output. Kerosene is also the fuel used in the aviation gas turbine, which has superseded the piston engine in all but the smallest aircraft. The fuel has to satisfy special requirements which are not relevant for ordinary kerosene. The most important of these is thermal stability, which ensures that the fuel can be heated to relatively high temperatures by air friction without leaving gummy deposits in the fuel system, and of course without becoming dangerously inflammable. Fuel meeting the requirements of civil aviation is obtained by distillation in the range 150–250°C, with perhaps some removal of aromatics, and it is referred to as aviation turbine kerosene or Avtur. There is another aircraft fuel in common use, known as aviation turbine gasoline or JP–4, which is obtained by distillation over a wider temperature range, usually 30–260°C. The wider cut enables more fuel to be produced from a given batch of crude oil, and during the oil shortages of 1974 it became common practice to use JP–4 instead of Avtur. However, the presence of low-boiling fractions increases the risk of fire in an accident, and JP–4 is generally discouraged in the United Kingdom.

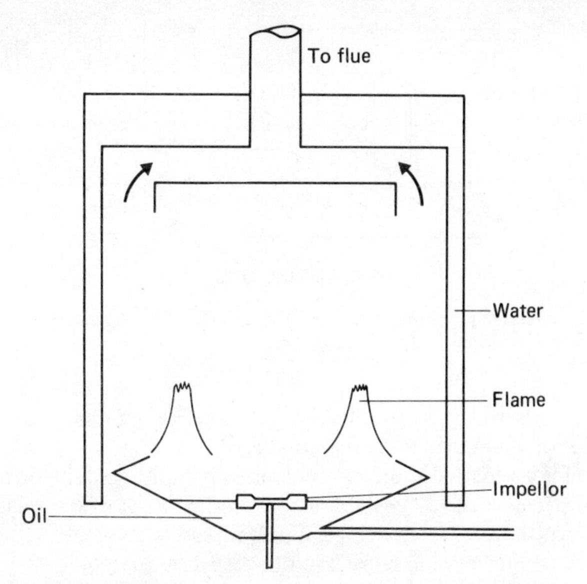

Fig. 5.10 Wall-flame burner. (From McMullan et al 1976. Reproduced by permission of John Wiley and Sons Ltd.)

Naphtha

The fraction boiling between 80 and 180°C, known as naphtha, is not used as such, but is processed by cracking into several different products. In the USA, where the demand for gasoline is higher than can be achieved by distillation alone, naphtha provides an additional source. In Europe, where the demand for gasoline is relatively lower, naphtha is cracked into gases such as ethylene, which are used as feedstocks for the petrochemical industry. It has also been used as a source of town gas as an alternative to coal gas. Neither of these uses has been necessary in America, for traditionally both the feedstock and the fuel gas markets have been supplied by natural gas.

Gas Oil

This is obtained by distillation in the range 250–350°C. As a heating oil it can be handled in similar types of burners to those used for kerosene, but it is now customary to use pressure-jet burners in which the oil is atomised by spraying through a fine jet under pressure (Fig. 5.11). Its other main use is as a fuel for diesel engines, which differ from petrol engines in having no spark, relying instead on compression to provide sufficient heat to ignite the fuel when it is sprayed into the cylinder. Clearly a diesel engine must have a very high compression ratio, and this gives it a higher efficiency than the petrol engine, which reduces operating costs. However, the engine must be

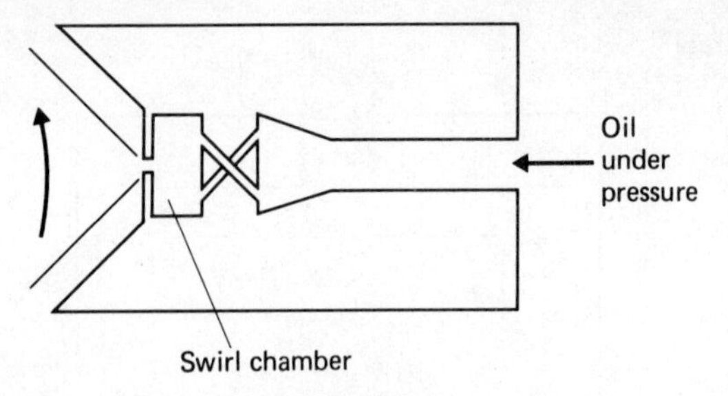

Fig. 5.11 Pressure-jet burner. (From McMullan et al 1976. Reproduced by permission of John Wiley and Sons Ltd.)

made stronger to withstand the high pressures and its capital cost is higher than that of a comparable petrol engine.

The characteristics of diesel fuel must obviously be quite different from those of petrol. Volatility is not needed, in fact it may be a nuisance. Spontaneous ignition, far from being a disadvantage, is actually a requirement. The performance of a diesel fuel in this respect is measured by its cetane number, which is analogous to the octane number of petrol. The highest cetane numbers are reached by the paraffinic components, but these tend

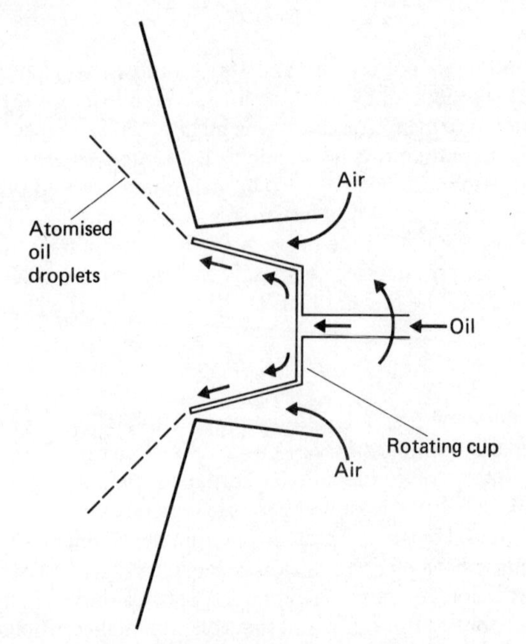

Fig. 5.12 Rotating-cup burner.

to solidify into waxes at low temperatures, and in an unexpectedly hard winter this can lead to an embarrassing number of breakdowns. Diesel fuels intended for use in cold weather are generally 'winterised' by various methods to prevent this occurrence.

Residual fuel oil

After all the distillates have been driven off, the remaining heavy oil is suitable only as a fuel for large-scale furnaces. The viscosity is so high at ordinary temperatures that it is necessary to use heated tanks and fuel pipes to enable the oil to flow at a reasonable speed. The main limitation in specifications of quality is the sulphur content, for excessive amounts of sulphur dioxide in the combustion products leads to air pollution and may damage the plant. Although some sulphur removal may be necessary, it is usual to control the sulphur content by blending with oil from low-sulphur crudes. The design of burners for heavy oil concentrates on the need for atomisation. The heated oil is atomised into droplets of less than 0.1 mm diameter by blowing under pressure from a nozzle or by means of a blast of air or steam, or by being thrown off a rotating cup. The droplets are blown into the combustion zone in an air stream and are ignited by means of a pilot flame. A typical burner is shown in Fig. 5.12. The most important use for residual oil is in electricity generation. The plant in an oil-fired power station is essentially similar to that in a coal-burning installation, the only major difference being the lack of ash disposal facilities. The siting of the station is different, however. The most favourable site is near an oil refinery because the fuel supply presents no problems, and as the refinery is already a major industrial installation, the addition of a power station causes no difficult environmental objections.

6 Nuclear Power

After fossil fuels, the other dominant source of energy in general use today is nuclear power. When we talk about nuclear power, what we really mean is the conversion of the energy locked up in the nucleus, through nuclear reactions, first into heat and then into electricity. It is electricity that is delivered to the consumer. In all present nuclear power plants, the energy is released through the breaking up of large nuclei, that is through nuclear fission. It is hoped that in the future, the power will come from the joining together of very small nuclei in the process called fusion, but this stage is still some way off.

Before examining either of these processes it is worthwhile looking at the basic physics of the nucleus. The atom is seen today as a very small, massive, and positively charged nucleus surrounded by light negatively charged electrons. It is these electrons that exercise control over both chemical reactions and the physical properties of materials. The effect of the nucleus is limited to determining the number of electrons that the atom will contain and the atomic mass.

The nucleus is pictured as being a conglomerate of positively charged, heavy *protons* and electrically neutral *neutrons*, which have a slightly greater mass than the protons. These *nucleons* are held together by so-called nuclear forces which act only at extremely short range and which are strong enough to overcome the electrostatic repulsion between the similarly charged protons.

It is possible for two nuclei to exist with the same number of protons but with different numbers of neutrons. Since the chemical properties of an atom are determined by the number of electrons, and hence the number of protons, these two different nuclei are chemically identical. They are called *isotopes,* that is they have the same *atomic number* (number of *nucleons*). However, because of this different nuclear composition these two atoms may have very different nuclear characteristics. A notation has evolved to denote on the chemical symbol both the nuclear charge, (atomic number, Z) and the nuclear mass (mass number, A). The chemical symbol is written down, the mass number is written as a superscript and the atomic number is written as a subscript. Alternatively, since the chemical symbol uniquely defines the atomic number, the subscript is frequently omitted. For example the isotope of carbon which contains six protons and six neutrons would be written as one of C^{12}, C_6^{12}, ^{12}C, or $_6C^{12}$. All are used, but the first and last are preferred.

One of the things that is apparent in a study of the properties of nuclei is that the ratio of neutrons to protons in the nucleus must lie within certain specified limits if the nucleus is to be stable. If the ratio is too large the nucleus will be radioactive and will decay in such a way that the ratio is improved. This can be either by gaining positive charge or by losing neutrons. On the other hand if the ratio is too small, that is, there are too many protons for stability, then the nucleus tries to lose positive charge to restore the ratio. This is illustrated in Fig. 6.1 for isotopes of carbon.

The ratio is not constant. As the atomic number increases, the nucleus requires a greater proportion of neutrons to remain stable. That this is not unreasonable can be seen from very simple arguments. If one tries to make a large nucleus with very few neutrons, then the repulsive electrostatic forces between the protons become enormous. The lack of neutrons means that the protons come closer and closer together as more are added. Thus the forces trying to disrupt the nucleus become greater. If a sufficiently large numbei of neutrons are included within the nucleus, however, they prevent the protons from coming too close and help to keep the repulsive

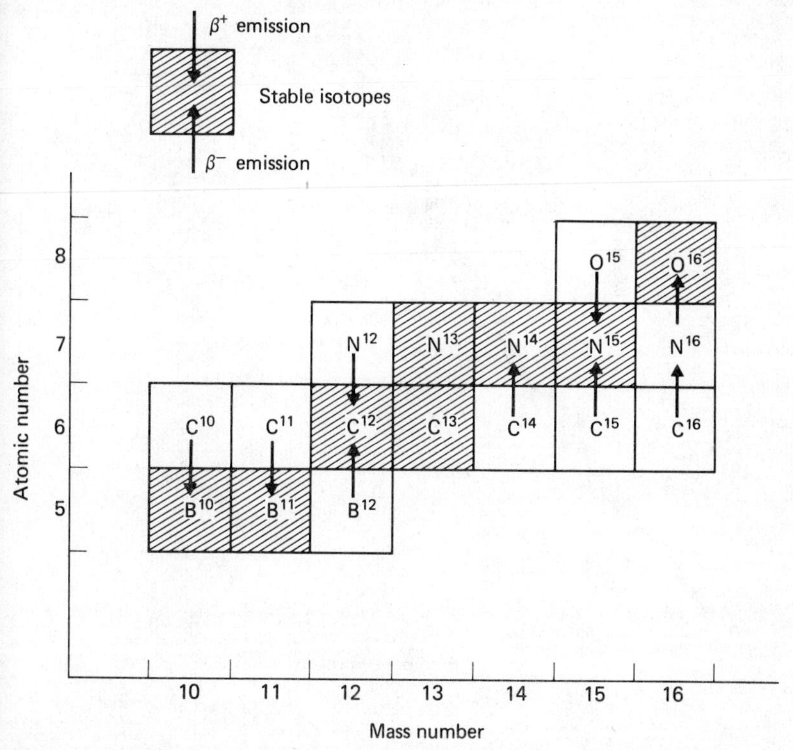

Fig. 6.1 Modes of decay of some isotopes of boron, carbon, nitrogen and oxygen.

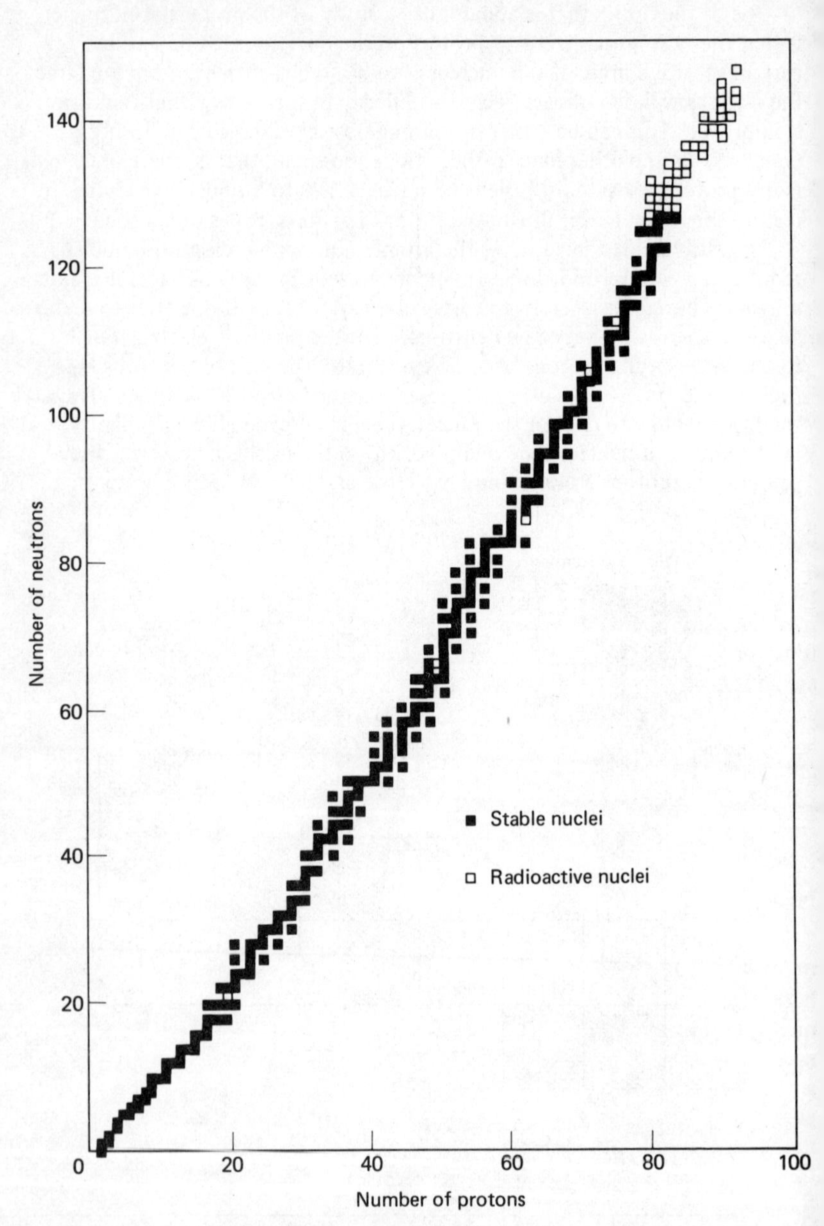

Fig. 6.2 Chart of the naturally occurring nuclei.

forces down to tolerable levels. The argument as to why an excess of neutrons cannot be tolerated is rather more complicated and depends on such factors as the nucleus adopting its minimum energy configuration (rather like a ball placed halfway up a hill rolling down to the bottom) and the distribution of energy between neutrons and protons within the nucleus. Any surplus neutrons will effectively decay to a proton and an electron and the electron will be emitted from the nucleus as a *beta particle*. However it is found that as the number of protons increases, the allowed number of neutrons increases rather more quickly. This is clear from Fig. 6.2 which shows the distribution of nucleons in the stable nuclides.

It is also apparent from Fig. 6.2 that there are no stable nuclides above a certain size. That is, if a nucleus is too heavy (has too many protons and neutrons, even if these are distributed in the proper ratio), then it will be unstable in the same way that a liquid drop becomes unstable if it becomes too large.

The other important idea for understanding nuclear power is the concept of binding energy. The more tightly bound a particle is, the more energy has to be supplied to it to free it. In other words, its *binding energy* is greater. It is found that in very light nuclei the binding energy of each nucleon is small, while in heavy nuclei it is greater. But there is a range of intermediate atomic numbers for which the binding energy per nucleon is greatest, as shown in Fig. 6.3. The result is that if two very light nuclei are fused together, the resulting nucleus has a mass less than the sum of the components and energy is released in the reaction. Also, if a heavy nucleus splits in two, the two components together have less mass than the parent, and energy is again released in the process. It is this feature that makes possible the use of nuclear fission and fusion in generating power, and incidentally in generating explosions.

Nuclear Fission

The discovery of nuclear fission was one of the results of attempts to produce elements with atomic numbers greater than 92 (uranium) by neutron bombardment. It was hoped that the neutrons would strike the target nucleus and combine with it to produce a nucleus with the same atomic number but with a greater mass number. This nucleus would be unstable, and would decay by beta emission to become a daughter nucleus with mass greater than the original target nucleus and also with a greater atomic number. These experiments eventually succeeded but the interpretation of the early results was difficult because of some observations which did not fit the expected patterns. It was found that these could only be explained if the target nucleus had broken into smaller fragments, and, in 1939, Hahn and Strassman discovered that when uranium was irradiated with neutrons, alkaline earth metals were produced.

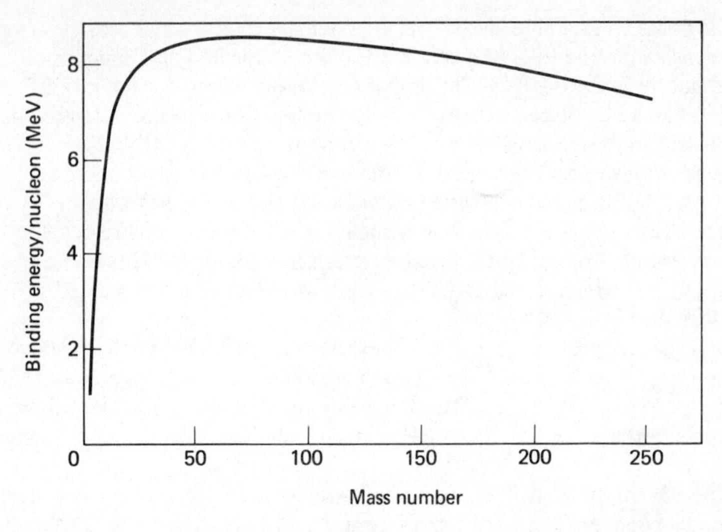

Fig. 6.3 Variation of binding energy per nucleon with nuclear mass.

In the same year, Meitner and Frisch suggested that, on absorption of a neutron, the uranium nucleus becomes sufficiently excited to break into two roughly equal fragments. Bohr and Wheeler, also in 1939, worked out a theory of the fission process based on a liquid drop model for the nucleus. They assumed that the nucleus behaves like a liquid drop and is held together by surface tension. This simple model is able to account for the instability of U^{235} and Pu^{239}, and the stability of U^{238}, when bombarded with slow neutrons, and shows how the heavy parent nucleus can break into two roughly equal fragments with the emission of large amounts of energy. In this *fission process* some 200 MeV of energy is released. Most of this goes into the kinetic energy of the fast moving fission fragments but about 20% is taken up by gamma rays, fission neutrons, beta particles and neutrinos. The uranium–235 reaction can therefore be written as

$$U^{235} + \left(\begin{smallmatrix} n \\ \text{slow} \end{smallmatrix}\right) \rightarrow U^{236} \rightarrow X + Y + \nu n + 200 \text{ MeV}.$$

X and Y stand for two intermediate mass fission fragments, and ν is the average number of neutrons released for each process. For the slow neutron–U^{235} reaction above, $\nu = 2.47$.

In the fission nuclear bomb, as many as possible of these fission neutrons are caused to strike another uranium nucleus, so producing a chain reaction which avalanches very quickly to produce an explosion. In the nuclear reactor, on the other hand, this process would be very embarrassing and steps are taken to absorb excess neutrons to ensure that only one fission reaction results from the neutrons of each earlier reaction. A necessary condition that a chain reaction can occur at all is clearly that $\nu > 1$. It is also

Fig. 6.4 The Steam Generating Heavy Water Reactor at Winfrith, Dorset. (Reproduced by permission of United Kingdom Atomic Energy Authority.)

necessary that the size of the sample of uranium must be large enough that the neutrons do not escape in large numbers before they can strike another nucleus. This is the reason why there is a critical mass of uranium below which neither a bomb nor a reactor is possible.

In the bomb, two subcritical masses of U^{235} or Pu^{239} are brought together to make a sample larger than the critical size. The build up of neutron intensity proceeds exponentially with a time constant of about 10^{-8} s. In less than a microsecond, therefore, a large fraction of the available energy is expended. The fission products, which carry most of the released energy as kinetic energy, are rapidly slowed down in the bomb material, giving up their kinetic energy as heat, so producing an explosion. The radioactive side products of the nuclear bomb are secondary to the explosion itself. Let us now turn our attention to more peaceful pursuits.

Nuclear Reactors

Each fission in U^{235} releases 200 MeV of energy. Since 1 MeV = 1.6×10^{-13} J = 1.6×10^{-13} Ws we see that each fission frees 3.2×10^{-11} Ws and 3.1×10^{10} fissions per second would release one watt of power. The complete fission of 1 kg of U^{235} spread over a twenty four hour period would produce energy in the form of heat at a rate of 1000 MW. If this heat can be turned into electricity with an efficiency of 30%, then the electrical energy would be supplied at the rate of 300 MW. This output is equivalent to that from a large power plant consuming 2500 tonnes of coal per day. It is this equivalence of 1 kg of a fissionable material such as U^{235} to 2500 tonnes of coal as a source of electric power that originally made nuclear power attractive.

Unfortunately, U^{235}, the isotope of uranium that readily undergoes fission, is the rarer isotope, being only 0.7% of natural uranium. Most of the rest is U^{238}, an isotope for which the probability of fission is very small but for which the probability is very high that neutrons slowing down to thermal energies by collisions in the reactor may be absorbed. It would be uneconomic to use pure U^{235} for commercial power production and so it is necessary to find ways of using natural uranium, or slightly enriched uranium, as the fuel in a reaction based on the fission of U^{235}. The answer is to slow down the neutrons as quickly as possible so that there is the least possible chance of the unwanted reactions with U^{238} taking place. This is done by including a *moderator* in the core of the reactor.

A moderator is a material in which the fission neutrons can be slowed down to thermal energies (that is to speeds of about 2200 m s^{-1}) outside the masses of uranium where the reaction is to go on. In order that a moderator should be efficient, it should absorb the maximum amount of energy from the neutron at each collision. This means that the atoms of the moderator should be as light as possible – ideally they should have the same mass as the neutron. This suggests that water might be a good moderator, which it is, but unfortunately hydrogen is also an extremely good absorber of thermal neutrons, so that water cannot be used as a moderator for natural uranium reactors. Heavy water, in which the hydrogen has already absorbed the neutron to form deuterium, can be used. Another compromise is to use carbon as the moderator. Carbon is light, common, inexpensive, and easily worked.

The nuclear fission reactor is possible because more neutrons are produced in fission than are needed to initiate it. As a result, a nuclear reactor can be built provided that it is possible to arrange for there to be slightly more than one thermal neutron produced – after all losses – to induce new fissions from each thermal neutron absorbed in the previous generation of fissions. It is possible to produce a sort of profit and loss account for the neutrons in a reactor which examines the creation and destruction of the neutrons within the reactor core. For example, suppose each fission by a thermal neutron produces n fast neutrons. Some of these will themselves induce fission producing a so-called fast fission fraction ϵ. In slowing down in the moderator, a fraction p of the $n\epsilon$ neutrons now available will escape being absorbed by uranium-238, and a fraction f will escape being absorbed in the structural material of the reactor or in the moderator while the neutrons have thermal energies. There will be other losses, for example neutrons may well escape from the core simply because it has a finite size. However, if the core is large enough these losses can be neglected, and we end up with a very simple expression; that each thermal neutron inducing fission will create k new thermal neutrons where

$$k = n\epsilon pf$$

This is the *four factor formula* and is very important in the design of reactors. Typically for a natural uranium reactor, $n = 1.33$, $\epsilon = 1.02$, $p = 0.9$ and $f = 0.9$, so that a typical value for k is 1.1.

When the reactor is running at its desired rate, some way must be found of reducing the factor k to 1.0, otherwise the number of fissions will steadily increase, the power generated will increase correspondingly and eventually the core will melt. This is done by using *control rods,* rods of some material such as cadmium or boron which has a high absorption for thermal neutrons and so can be used to soak up any excess neutrons and adjust k to any desired value.

We can now see a possible structure for a nuclear reactor as a large block of moderator into which is inserted the uranium fuel (usually as rods for ease of handling). Between the fuel rods would be inserted the control rods, which can be moved in and out according to the value of k needed at any time. This is shown in Fig. 6.5. Very many different types of reactors have been built and are in operation throughout the world. Natural uranium reactors have been built with both heavy water and graphite moderators. Light water moderated reactors have been built using as fuel uranium enriched in U^{235} to compensate for the extra absorption of neutrons by hydrogen.

The absorption of neutrons by U^{238}, which we have been trying to avoid, can be turned to advantage by using the reaction

$$_{92}U^{238} + {}_0n^1 \rightarrow {}_{92}U^{239} \rightarrow {}_{93}Np^{239} + {}_{-1}\beta^0$$
$$\searrow {}_{94}Pu^{239} + {}_{-1}\beta^0$$

to produce an isotope of plutonium, Pu^{239}. This isotope of plutonium has a high probability of undergoing fission when bombarded by slow neutrons and so represents an addition to the fuel content. As a consequence, the loss of neutrons by absorption in U^{238} is actually a gain as far as overall fuel economy is concerned. In fact, in the process of burning the comparatively rare U^{235}, the natural uranium reactor produces new fuel from the apparently useless U^{238}. This is the basis of the breeder reactor in which more nuclear fuel is produced than is consumed.

This, then, is the background to the development of nuclear fission power stations. The world's first commercial nuclear power station was opened at Calder Hall in the United Kingdom, in 1956. This station produced 50 MW of electricity. Since then, the number of nuclear stations has steadily increased throughout the world and the newest of these are to generate 2000 MW.

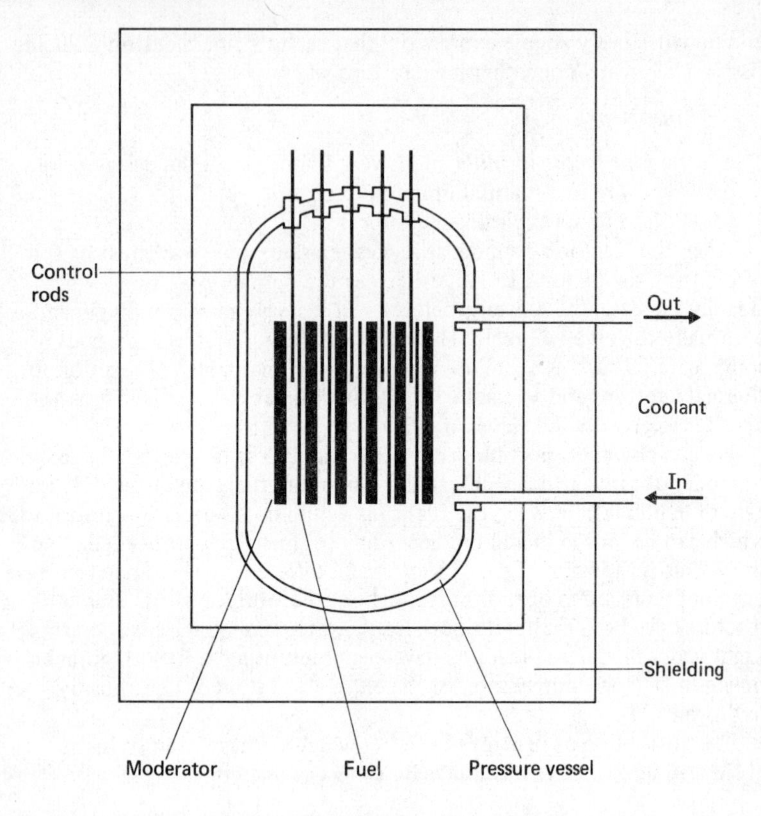

Fig. 6.5 Schematic structure of a nuclear fission reactor.

Magnox, AGR, HTGR, LWR, SGHWR, LMFBR, DFR, PFR! Not only has the nuclear power industry produced engineering and social problems to be solved as part of its implementation, but it has inspired a range of linguistic tortures as well. All these acronyms refer to different types of nuclear power stations. For example Magnox is the name given to the first British commercial stations in which the fuel was natural uranium contained in MAGNesium OXide alloy. They are gas cooled, using carbon dioxide, and the moderator is graphite. AGR is the name given to the second generation of British reactors, the Advanced Gas-cooled Reactors, and so on.

Cheap power must be the ultimate aim of any power generating system, though many other factors, such as air pollution or radioactive waste disposal problems, must be taken into account. Furthermore, in the present fuel situation, planning has to be done to extend our power production for the longest possible time using the known fuel types, while other alternatives are investigated. For this reason alone, nuclear power offers the best solution as fission fuel reserves are potentially greater than fossil fuel reserves.

There are two main types of reactor in common use in nuclear power plants today. These are the Light Water Reactor (LWR) which is widely used in the United States and which has come in for much criticism, and the gas cooled type which has been favoured in the United Kingdom. Other reactor types are used, for example CANDU, the Canadian heavy water reactor, but these other models are less common at the moment. As the name suggests, the light water reactor uses ordinary water, H_2O, as the moderator and as the coolant. Because of the way in which the proton, the hydrogen nucleus, easily absorbs slow neutrons, the uranium fuel has to be enriched in order that the chain reaction can be self sustaining. Indeed, it is one of the criticisms levelled against the design that the cost of enrichment for the first charge of fuel can be as much as 10% of the capital cost of the power station. Nevertheless, this is a very popular design.

There are two basic types of light water reactor, the *boiling water reactor* (BWR), and the *pressurised water reactor* (PWR). In the BWR, the cooling water boils in the core, and the steam generated is used directly to drive a steam turbine, which drives the generator. The steam is then condensed to water and pumped back to the reactor to complete the cycle. Thus, the reactor is the boiler for the whole process. In the PWR on the other hand, the core cooling water is kept at a very high pressure and is heated to some 600°C. It is then sent to a separate heat exchanger where a secondary water supply is boiled and used to drive the turbine. The two types are contrasted in Figs. 6.6 and 6.7. The problem with the boiling water type is that the cooling water, circulating through the core, becomes radioactive from slight leaks in the thin cladding of the fuel rods, and from radioactivity induced by neutrons just outside the cladding. Thus radioactive steam goes directly to the turbine so great care must be exercised to avoid steam leaks in the turbine itself. This difficulty is avoided in the pressurised water system as the cooling water and the steam for driving the turbine are kept separate. Much of the early experience with the pressurised water reactors was gained from nuclear powered submarines, where the nuclear engine, which does not need air for burning, was ideally suited. In addition, cost was no real limitation.

The light water moderated reactor has come under considerable criticism in recent years. Two of the main criticisms are very important indeed. Firstly, it is alleged that the technology of welding the very heavy steel sheets of the pressure vessel is not capable of providing the necessary reliability. This is especially important because of the catastrophic nature of the accident that would occur if the pressure vessel were to rupture. Against this the manufacturers claim that the chances of this happening with the present technology are so small that the risk is acceptable. Secondly, there are the possible effects of a sudden failure in the water supply to the core of the reactor, because of a burst or blocked pipe for

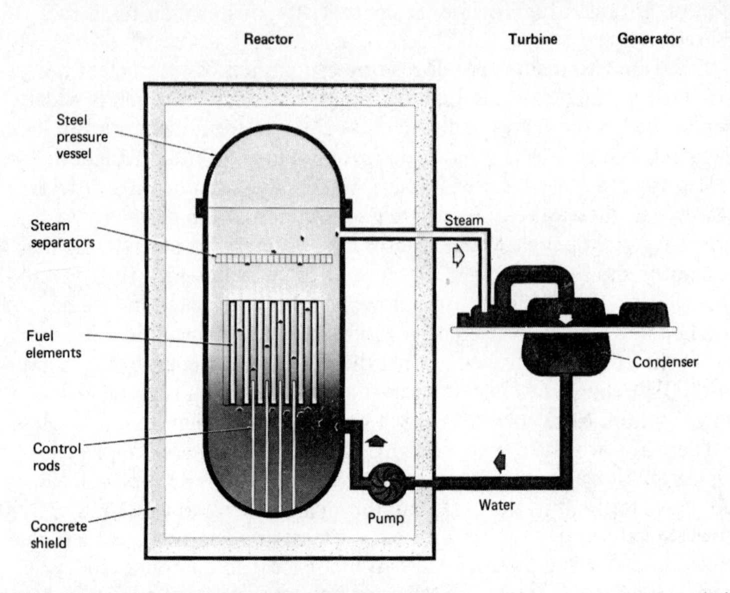

Fig. 6.6 Schematic diagram of the Boiling Water Reactor. (Reproduced by permission of United Kingdom Atomic Energy Authority.)

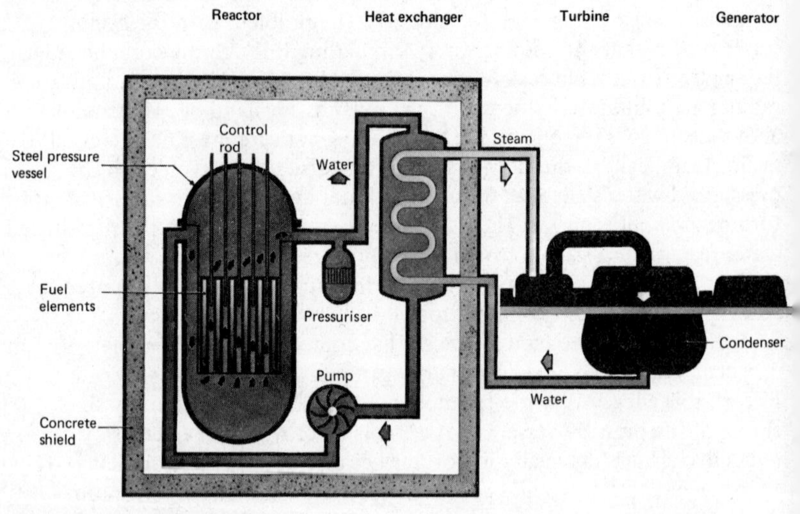

Fig. 6.7 Schematic diagram of the Pressurised Water Reactor. (Reproduced by permission of United Kingdom Atomic Energy Authority.)

instance. If this were to happen, goes the criticism, the large mass of fuel, and radioactive fission products, could become so hot as to melt, fall to the bottom of the containment vessel, melt through, and escape into the ground beneath. From here the radioactive contaminants could possibly infiltrate the ground water supply and become a hazard to life for a wide region around the reactor. This criticism is especially levelled against the PWR because water is an extremely good moderator so that the core is tightly packed. The BWR is moderated partly by steam and so occupies more volume. Emergency water supplies have been designed and installed in all water cooled reactors to combat this type of failure but, say the critics, since the failure has never occurred, the safety system has never been tested, and so cannot be assumed to work.

In fact, the failures that have occurred in the water cooled commercial reactors have all been minor, such things as sticking valves, steam leaks in the secondary system and so forth. This is not to say that serious accidents may not happen, only that they have not happened so far in something like 100 reactor years of experience with this type. It should be emphasised, however, that, in commercial terms, 100 reactor years is not a very long time.

The other design of reactor in large-scale use in commercial power stations, and the one which has been favoured by the British authorities, is the gas cooled type exemplified by the Magnox design in Fig. 6.8. In this design, the reactor is moderated by graphite, and the cooling is achieved by

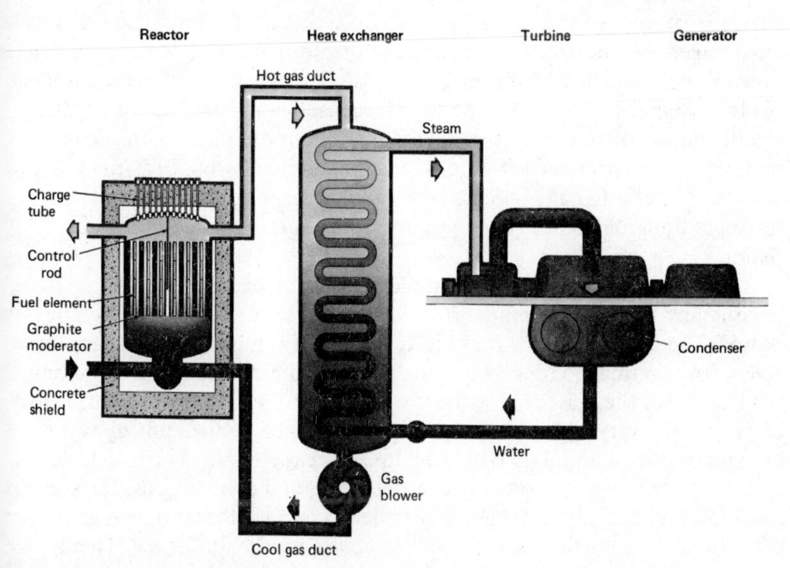

Fig. 6.8 Schematic diagram of the Magnox Reactor. (Reproduced by permission of United Kingdom Atomic Energy Authority.)

passing large volumes of carbon dioxide through the pile. The hot gas is then passed through a heat exchanger to produce the steam for the turbines in a secondary circuit. One design difference between the two types is that the gas cooled reactor is physically much bigger than the water cooled type. This has been quoted as a safety factor because the core is more dispersed, and there is less likelihood that a cut off in the coolant supply will cause the core to melt. Also since the graphite expands as it becomes hotter, there is a tendency for the fuel density to become lower, making the reactor less critical. This means that less heat is generated and so there is a certain amount of self-damping.

In practice, there is not just one heat exchanger per reactor with either type of unit. Instead, the coolant from the reactor core is passed to several different heat exchangers. This simplifies the engineering problems and improves the cooling efficiency. In addition it reduces the risk of accident through the failure of one single element.

The Breeder Reactor

As was pointed out earlier, all reactors running on either natural or enriched uranium as fuel lose some of their neutrons through absorption by U^{238}. In the usual commercial reactors this represents an immediate loss of neutrons though the resulting plutonium does add to the total fuel content of the reactor. This increase in fuel is less beneficial than the loss of neutrons is harmful and so steps are taken to minimise it. In the *breeder reactor*, however, the capture of the faster neutrons by U^{238} to form Pu^{239} is positively encouraged. No moderator is used in the reactor core to slow the neutrons down rapidly and as a result the capture of neutrons by U^{238}, which is most likely at higher speeds, is enhanced. As a consequence the reactor produces significant quantities of plutonium and, in fact, more plutonium fissile material can be produced from the otherwise embarrassing U^{238} than there was Pu^{239} fuel originally – hence the name breeder reactor. It represents the most hopeful chance of long term power production from nuclear fission.

The breeder reactor has some unpleasant characteristics which are regarded by its critics as rendering it unacceptable for generating electric power. The first of these is that plutonium itself is highly toxic. It also has a very low thermal conductivity which adds to the difficulty of extracting the heat from the reactor core. Further, because there is no moderator, the core runs at a very high energy density and must be cooled, not by water or a gas, but by a liquid metal – sodium. Since the relatively small core volume is taken up by closely packed fuel rods and control rods, the sodium must reach extremely high speeds in order to remove the heat as it is generated. Consequently it is subject to considerable turbulence. Therefore, goes the argument, any impediment to its flow could lead to the failure to remove 400 kW of heat from each litre of reactor volume. This situation would lead to a melt down of the reactor core if left uncorrected. Thus one of the criticisms of the breeder reactor is almost identical to the corresponding criticism of the light water reactor. Even the coolant itself is subject to criticism. Sodium is chemically a highly reactive material. In particular, it reacts explosively with water, as many schoolchildren can testify. In the breeder reactor, there is a large quantity of sodium, at an elevated temperature, and therefore more reactive, being pumped around the reactor core. Not only is the coolant highly chemically reactive, but after a while, it is highly radioactive. A burst pipe or some similar leak in the cooling system could have extremely nasty consequences.

The other major criticism of the breeder reactor is rather less a criticism of the reactor itself than of the society in which we live. The breeder reactor uses plutonium as its primary fuel. The fuel rods are enriched in Pu^{239}, and the whole system is designed to produce more Pu^{239} from the U^{238} which is included in the reactor. One can make a nuclear bomb from the Pu^{239} which the breeder burns and makes. Any aspiring nuclear bomb maker can therefore acquire the weapons grade plutonium (greater than 90% Pu^{239}) that he needs from the breeder reactors. The plutonium in the wastes from conventional reactors varies from between 70% Pu^{239} in the light water reactors to about 85% in the Magnox reactors, and since this is contaminated with other plutonium isotopes, it is not practicable to separate out the Pu^{239}. By comparison, the wastes from the breeder contain plutonium that is already over 90% Pu^{239}! One estimate has it that all our bomber needs is one graduate physicist, some technicians, at most $100 000 worth of standard laboratory equipment, six months peace and quiet – if one will pardon the expression – and some fissile material. It is this one special ingredient that the breeders can supply.

Notwithstanding these enormous qualifications, it is still true that the breeder reactor represents the only way of extending nuclear fission fuel reserves significantly beyond those of coal and oil. It is a question for society to answer whether or not the price is worth paying.

Reactor Safety

In the above discussion, we have seen some of the risks associated with nuclear reactor power production. However, the actual record of the industry is good. The accident rate is low and those accidents that have occurred have generally been minor – and of the type that occur in traditional power stations without any public outcry. Both the power generating companies and the designers and builders of nuclear power stations recognise quite clearly that the nuclear fission reactor carries its own peculiar hazards. They go to great lengths to protect the public and also the power station personnel from these hazards. It must be pointed out, however, that this does not mean necessarily that they have been successful. In the words of Professor Hannes van Alfvén, a 1970 Nobel prize winner in Physics,

> 'The reactor constructors claim that they have devoted more effort to the safety problems than any other technologists have. This is true. From the beginning they have devoted much attention to safety and they have been remarkably clever in devising safety precautions. This is perhaps pathetic, but it is not relevant. If a problem is too difficult to solve, one cannot claim that it is solved by pointing to all the efforts made to solve it.'

What is the record so far? There have been accidents, though the accident record is better than for the rest of industry. A study of some of the early ones showed that they were typically due to one or more of three factors; carelessness, design error or mechanical failure. The accidents of the past, of course, help to design the safety systems of the future, and so while unfortunate, they are not entirely negative in their effect. We will illustrate the point by looking at three accidents that have occurred.

The first of these occurred at Chalk River, Ontario in 1952. The reactor was a low power heavy water reactor, with ordinary light water coolant circulating in pipes through the moderator. This type of reactor is inherently unstable in the event of a cooling water failure. The light water absorbs some neutrons, and this loss is included in the reactor calculations. If the cooling water supply fails, the absorption of neutrons decreases, so that more are available for causing fission – increasing the activity of the reactor. This is what happened at Chalk River. An operator turned some valves in the cooling water circuit, selecting the wrong ones. The chief operator ran to see what had happened and to adjust the valves. He asked for the control rods to be inserted into the core to stop the chain reaction. Unfortunately the wrong buttons were pressed and by the time this error was corrected the core was too hot and the control rods would not work. In a few seconds, the sudden rise in activity melted the uranium which reacted with the remaining water to form hydrogen gas. The core exploded. Safety devices operated, however, which prevented radiation injury to people in nearby laboratories.

The second of our three accidents took place at Windscale in the United Kingdom in 1957. The reactor was a plutonium producer and was essentially a large gas-cooled cube of graphite containing uranium. This was an accident of a much more technical nature. The neutrons in the core of the reactor are slowed down by collisions with the atoms of carbon in the graphite. In the collisions energy is given to the carbon atoms and many of them are knocked out of their preferred sites in the graphite structure. Because the graphite would prefer to be left alone, this creates additional internal stresses, and stores extra energy in the graphite. If these stresses are not relieved slowly and regularly they build up and eventually the core will explode. The way of relieving the stresses is a process called annealing in which the temperature of the graphite is raised slightly, so giving the atoms trapped in the wrong places just enough energy to move, when most of them find their way back to their preferred sites in the graphite lattice. At Windscale the routine procedure for doing this went wrong and a fire of molten uranium and carbon burned in one part of the reactor for a day or two before the trouble was recognised and the fire extinguished. Radioactivity from this accident made it necessary to dump milk supplies from grazing land for several miles around the reactor, and higher levels of radioactivity were also found in parts of the Irish Sea.

The last of these accidents is noteworthy primarily because it happened on a prototype of the commercial fast breeder reactors. Because these reactors are compact and there is no delay time for slowing down the neutrons, they are more sensitive to control adjustments, and the demands on the integrity of materials also tend to be increased. Near Detroit in 1966 an accident happened in the Fermi fast breeder reactor. This reactor was designed to operate at between 200 and 300 MW. It had been shut down and was being restarted again. Suddenly, when running at about one-tenth of full power, the meters started showing that something was wrong. The reactor quickly heated up and excessive radiation was detected by the monitoring systems which released the emergency controls and the reactor was shut down. In the true sense this was not an accident. The safety systems worked, the reactor was shut down, the radiation was almost entirely kept within the outer containment shell, and there were no serious external effects. It was not so much the actual event that caused concern but what might have been. In the accident, some of the fuel rods melted and it was feared that there might be a dangerous near-critical lump of fuel inside the reactor. If this was so then the reactor could have become a sort of very slow, low power, nuclear bomb. Actually, 'bomb' is too strong a term as the time scale for the reaction would be too long to cause a large explosion. Nevertheless the resulting nuclear fizzle would not be pleasant.

One of the things that reactor builders have to show is that they provide protection against the 'maximum credible accident' – that is, against the largest possible malfunction that can be foreseen as happening in a particular reactor installation. In the case of the Fermi reactor, the maximum credible accident was foreseen as being the melt down of one fuel assembly, of which there were about one hundred in the core. In the actual accident three of these assemblies melted! Despite the happy ending, this incident exemplifies the need for concern over the dangers of unforeseeable accidents in nuclear power plants.

Radioactive Waste

While the possible results of an accident in a nuclear power plant are frightening enough, the implications for the future posed by the long term storage problems associated with the waste products are even worse.

Let us look again at the physics of nuclear fission. When the neutron strikes the U^{235} or Pu^{239} nucleus, amalgamates with it, and causes fission, it does so because the nucleus has too many neutrons compared to its number of protons, and is therefore unstable. The nucleus breaks into two unequal parts together with some fission neutrons. Fig. 6.2, the chart of the stable nuclei, shows that the ratio of neutrons to protons in the stable nuclei increases with the atomic number Z. Thus, both fission fragments have too many neutrons for stability at their atomic number (Fig. 6.9). They are therefore radioactive.

The fission fragments always occur in pairs, one heavy and one light, though the actual masses of the fragments are distributed over a range of mass numbers. Thus there is a range of possible atomic species that can occur in fission. Several of these species set up a chain of radioactive decay processes in their search for stability. For example, one common fission chain from a heavy fragment is the one at mass number 140,

$$Xe_{54} \xrightarrow{\beta^-} Cs_{55} \xrightarrow{\beta^-} Ba_{56} \xrightarrow{\beta^-} La_{57} \xrightarrow{\beta^-} Ce_{58} \text{ (stable).}$$

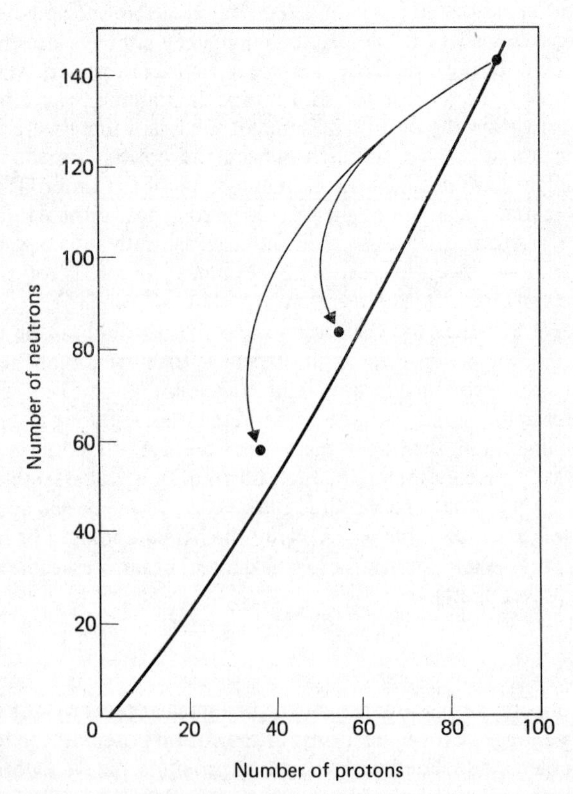

Fig. 6.9 The fission of U^{235} to produce two probable fission fragments. These are both above the stable isotope curve and are therefore radioactive.

This is the chain in which fission was discovered. It is among the most probable modes of fission of U^{235} and occurs in about 6% of fissions.

We have as a direct result of the fission process, therefore, many direct fission fragments, which are radioactive, and, in addition, chains of radioactive decays to other radioactive nuclei as part of the search for stability. Unfortunately some of these radioactive decay products are extremely long lived. For example, Sr^{90} and Cs^{137} have half-lives of 28 and 27 years respectively and so must be stored for over 500 years to reach the recommended 20 half-lives – that is to about one millionth of the original activity. Thus, the power we generate today builds up a legacy of problems for our children, grandchildren, great-grandchildren, ... The decision, therefore, to extend our nuclear energy programmes cannot be taken lightly. Nonetheless, since we will probably have to do so, the problems of waste 'disposal' must be examined.

In the past, while the amounts of radioactive waste to be taken care of have been fairly small, they have been stored in a variety of different ways: in large concrete tanks, either above or under the ground, in disused mines, in porous soil beds, in salt beds deep underground, and at the bottom of the oceanic trenches. The experiments into disposal in the salt beds were discontinued after fears became widespread that water might dissolve the salt and allow the radioactive waste to contaminate the ground water supply. Burial in porous soil beds was discontinued after it became apparent that in one particular case in the United States, Trench Z9, the soil bed was found to be acting rather like an enormous chromatograph column and was preferentially segregating out different radioactive waste products at different depths. It was feared that some soils were approaching critical density so that fears of a nuclear incident were rising. The AEC were given authorisation to dig it up again and remove the soil to a safer depository.

Other more esoteric methods of disposal have been suggested, including burial in the antarctic polar ice cap where the container would melt its way through the ice cap under the effect of its own radioactive heating, until it came to rest on the ground. Other suggestions are to send it into the sun in a rocket, or to bury it in the junction of two of the earth's tectonic plates so that it will be dragged down under the earth's mantle for a few million years. In view of the difficulties, however, it is probable that the concrete storage farms will remain as a long term feature of radioactive waste disposal.

Disposing of radioactive waste has several different aspects. Firstly there is the conventional problem of storing chemically active materials. Secondly there is the problem of radioactive containment, but this poses no great difficulties and really only means thicker walls to the containment vessels so as to attenuate the radioactive intensities outside the tanks to acceptable levels. Thirdly, however, there is the problem that the radioactivity releases energy into the body of the waste. This has to go somewhere and it does so by heating the waste material. Thus the radioactive waste has to be kept cool. This is the effect that leads to the suggestion of melting through the antarctic polar ice cap.

As the volumes of waste go steadily up and up, these problems will become more and more severe. The sizes of the tank farms will increase, as will the amounts of radioactivity being stored. Once again, like the breeder reactor, the possibility of terrorist activity arises and so there must be tight security on the waste farms. As time goes by this security problem will become greater and greater, and this is another aspect of nuclear power that must be borne in mind before any large-scale programme is developed. It is this problem that has prompted many people to suggest that, while power from nuclear fission may be necessary in the short term, it must be only an interim measure to be replaced as soon as possible by some less dangerous method of generating electricity.

Other Applications of Nuclear Energy

There is no doubt that the nuclear fission reactor driving a turbo generator is going to supply all the electricity to be gained from nuclear energy in the short term. Nevertheless there has been some work done on direct conversion of nuclear energy to electricity. In these reactors, the fuel rods are clad with a very thin layer of tungsten, and are held in position with a space between them and another electrical conductor. When the reactor is running it is allowed to become very hot, and the thin tungsten layer emits electrons by thermionic emission. These electrons flow across the space to the other conductor and so an electric current appears. This direct thermionic reactor has produced 20 kW of power for long periods.

Another suggestion has been to use nuclear power to make steel. This has been an intriguing possibility for several years but the problems of using a reactor to heat strongly reducing gases such as hydrogen and carbon monoxide to the necessary 1000°C has always inhibited its use. One suggestion that has appeared, however, is to use two reactors, one inside the other. The inner reactor is to be the latest High Temperature Reactor (HTR) in which the coolant is helium which is raised to 1200°C and then gives up its heat to the gases used in the iron ore reduction. Because of the containment problem with the HTR it is to be surrounded by a conventional Advanced Gas Cooled Reactor cooled by carbon dioxide at 500–600°C and raising steam for electricity generation in conventional reactor heat exchangers. The purpose of this complex proposal is to try to overcome the engineering problems of containing the high temperature high pressure helium coolant and also to produce good overall thermal efficiency. It remains to be seen if such a scheme is feasible.

Nature Got There First

Despite man's conviction that the nuclear reactor is a new device which is either the potential saviour of mankind or the cancer that will destroy us all, there is clear evidence that 1700 million years ago natural forces built a water cooled reactor that operated intermittently for as long as a million years. This phenomenon occurred in what is today Gabon in West Africa, and was revealed in 1972 by nuclear physicists of the French *Commisariat à l'Énergie Atomique.* The initial evidence was that, for the first time, uranium deposits were found that contained less than the standard 0.72% of the isotope U^{235}. These deposits were at Oklo in Gabon. Further detailed examination of several different batches showed some deposits with as little as 0.44% U^{235}, and also curiously enough some deposits showing slight enrichment to 0.74%. Previous to this discovery the standard 0.72% had been invariant. What could be the explanation?

The obvious though startling explanation is that at some time in the remote past, when less of the U^{235} would have decayed so that the natural ore was about 30% U^{235}, geological conditions had conspired to bring together sufficient uranium ore to become critical and form a nuclear reactor preferentially using up the U^{235}. However, what about the enriched deposits? These too could be accounted for because the fission of the U^{235} would produce large quantities of plutonium. Over the years this would decay back to uranium, and since this was younger than the original, it would be at a greater enrichment of U^{235}. This is all very well, but perhaps a little far-fetched. Fortunately the ores were also found to contain four particular rare elements: neodymium, samarium, europium and cerium, all with isotopic compositions only found previously in man-made nuclear reactions! The physical evidence was by now becoming extremely strong. All that was really necessary was to show that the geological conditions were suitable at some appropriate time in the past and the evidence for this natural nuclear reactor would be overwhelming.

The uranium bearing deposits in Gabon lie in a series of sedimentary rocks that are some 1700 million years old and which filled up great depressions in the granitic and crystalline basement of the region. The heavy rainfall of the area and other climatic conditions combined to determine the nature of the sedimentation and also the mineral accumulation. Sedimentation in the basin followed a sequential pattern in which conglomerates were covered by layers of finer and finer materials ranging from sandstone to silt. Mineralisation occurred during the early stages of this cycle. After deposition the sediments were subjected to powerful tectonic upheavals which induced transformations and helped to reconcentrate the uranium bearing ores which border a fault system running between the basin and crystalline rock formation. This is shown in Fig. 6.10.

We now have the situation where the uranium ores have been concentrated – even today the uranium concentrate in some of the ores from Oklo exceeds 10%. They lie in a water saturated layer of sandstone about 20 feet thick which is imprisoned beneath a layer of impermeable clay. In addition, the nature of the sedimentary deposits was such that the area was

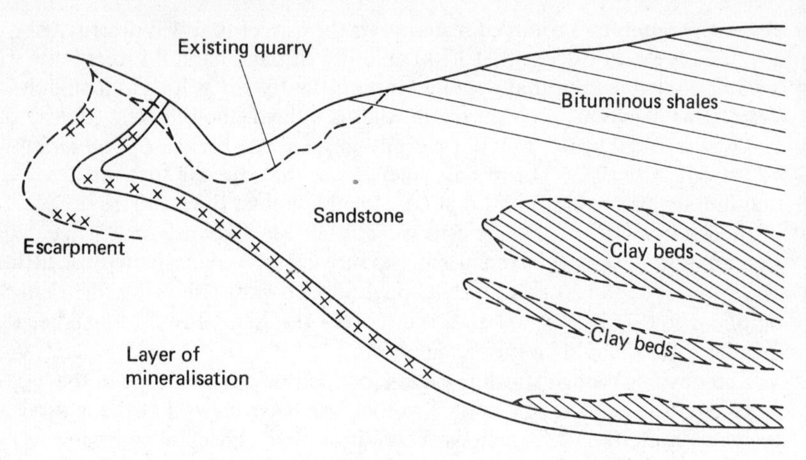

Fig. 6.10 Structure of the natural fission reactor at Oklo, Gabon.

almost totally devoid of neutron absorbing elements apart from the water. In fact nature had contrived to produce its own enriched light water moderated reactor. The indications are that, once the reaction started, it continued until the water had turned to steam when it stopped to allow cooling. Meanwhile the reaction broke out in another place, and so on for about a million years.

If this is an unique site then the consequences for the world supply of fissile uranium are slight, however unfortunate it may be for Gabon's export market. But, if the effect is more widespread, the consequences are much more serious and indeed buyers of uranium ore are now likely to insist on an 'isotopic guarantee' when placing forward orders now that the previously invariable 0.72% has been violated.

Nuclear Fusion

Nuclear fission is the most attainable method for the use of the energy stored in the nucleus, relying as it does on the near instability of extremely heavy nuclei, and the fact that the two component nuclei resulting from the fission have more neutrons than they would have in their stable configuration. At the other end of the nuclear mass spectrum, there is the related phenomenon of nuclear fusion, in which two very light nuclei are fused together to form a heavier one. Once again, this nucleus is heavier than its stable form and energy is released. If we look at the detailed structure of the binding energy curve for the very light nuclei, we find that the alpha particle, or helium nucleus, with two protons and two neutrons, has a particularly high binding energy per nucleon, that is, it is particularly stable.

The suggestion is clear. If it is possible to force two protons and two neutrons together, sufficiently close that the very short range nuclear forces can become dominant, then there is a good chance that they will coalesce to form helium and that energy will be released in the process. The most obvious way of driving these four particles together simultaneously is to find a half-way stage where part of the work has been done already and to start from there. This stage exists in the form of deuterium and tritium, two heavy isotopes of hydrogen, containing one and two neutrons respectively. When we look at the possible reactions, we find that there are four candidates.

$${}_1D^2 + {}_1D^2 \rightarrow {}_2He^3 + {}_0n^1 + 3.27 \text{ MeV (neutron branch)}$$

$${}_1D^2 + {}_1D^2 \rightarrow {}_1T^3 + {}_1H^1 + 4.03 \text{ MeV (proton branch)}$$

$${}_1D^2 + {}_1T^3 \rightarrow {}_2He^4 + {}_0n^1 + 17.6 \text{ MeV}$$

$${}_1D^2 + {}_2He^3 \rightarrow {}_2He^4 + {}_1H^1 + 18.3 \text{ MeV}$$

As 1 MeV is equal to 1.6×10^{-13} J, it follows that if 1 gram of deuterium and the corresponding weight of tritium (about 1.5 grams) are used in the third reaction, then the energy released will be approximately 85×10^{10} J. If this amount of energy were to be converted to electricity at an efficiency of one-third, and used in a period of one hour, then it corresponds to an electrical power of 78 MW. Thus 1 gram of deuterium per hour is potentially equivalent to 78 MW of electrical production. This is an extremely attractive proposition, being equivalent to burning 27 tonnes of coal per hour, but unfortunately there are several snags.

These reactions are the same in character as those that drive the sun. In the case of the sun, millions of years were needed for the gravitational attractions between the neutrons to pull them sufficiently close together for the nuclear reactions to start. On the earth we neither have access to sufficiently large numbers of neutrons to build a star, nor do we have sufficient time for such a project to evolve naturally. A quicker route must be found.

The principal basic difficulty is that, in order for the reaction to be possible, the nuclei must come very close together. Unfortunately, the nuclei are electrically charged. Consequently, as they come closer and closer together, there is a Coulomb repulsion between them that increases as the separation decreases. This force holds until the two nuclear centres are closer together than about 5×10^{-15} m. It is a straightforward matter to estimate the amount of energy that is needed to surmount this barrier in the case of the fusion of hydrogen nuclei. Using Coulomb's Law and some elementary ideas of thermal equilibrium, we can calculate the energy required as being about 4.5×10^{-14} J, and the corresponding temperature as about 2×10^9 K, that is, 2000 million degrees. This is an exceedingly hot gas.

Fortunately, this calculation is not accurate. While it does give an estimate of the temperatures required, it is limited in many respects. The most important limitation is that it suggests a precise temperature above which fusion can occur and below which it cannot. This is not observed in practice. In any actual experiment two phenomena will help to reduce the temperature at which the onset of fusion will be observed. The first is that if a gas is heated to some temperature, T, then not all the particles in the gas have the same energy. In fact, there is a continuous spread of energies present, ranging from particles that have been momentarily stopped in a collision, to some with extremely high energies after having experienced a number of 'lucky' collisions, each of which speeded them up. This is the Maxwellian distribution. Therefore, even in a gas at say 10^7 K there will be a number of particles with the 4.5×10^{-14} J which we have postulated for fusion. Further, it is not true that there is a definite onset energy. The reaction probability is much more complicated, and is seriously affected by a process known as *barrier penetration,* or *tunnelling.* This is a phenomenon well known in physics, whereby things that should not happen can. Basically, it means that even if a particle has an energy less than that needed for the reaction to proceed on a straightforward classical analysis, then there is still a chance, though a reduced one, that it will proceed. As the energy of the particle falls further and further below the threshold, then the likelihood of the reaction's proceeding falls off, rather than cuts off. The combination of these two effects is shown in Fig. 6.11 and shows that there is a strong probability that fusion reactions will be observed in a gas at temperatures as low as 2×10^7 K. In fact, at this temperature most of the fusion reactions will be occurring with particles having energies of 2×10^{-15} J, because of th influence of the Maxwell distribution. The term plasma is used to describe this hot mixture, because, at these temperatures, the electrons are stripped off the atoms, and we have a mixture of individual charged particles.

Thus, the only process reasonably open to us for attaining these temperatures is somehow to heat the gas. How can we do this, and what progress has been made? There has been considerable progress in laboratory and general experimental terms. Fusion reactions have been seen and measured in particle accelerator experiments. In these experiments, nuclei are accelerated through the appropriate electrical potential differences to give them the required energy and then impinge on a target of the other reacting material. In this way, important experimental data can be established, but these experiments canot lead to fusion reactors; the energy fed in is greater than would be extracted. On the other hand, the hydrogen bomb has been built and exploded. In this device, a conventional fission bomb is used to drive the deuterium–tritium mixture together with sufficient force to induce the fusion reaction. This then proceeds explosively. Thus, terrestrial fusion reactions are possible. Unfortunately, the hydrogen bomb is a little extreme as a power generating tool. The problem of taming nuclear fusion is that of finding some way in which the necessary tempera-

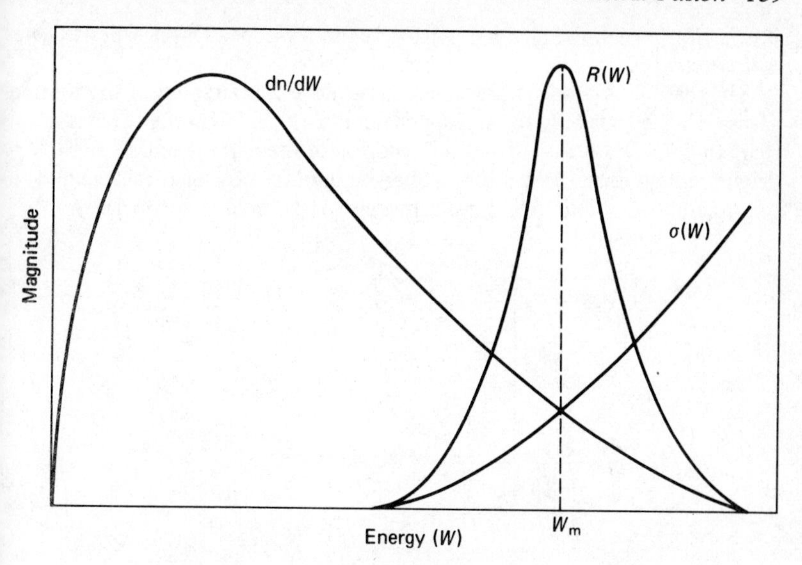

Fig. 6.11 Variation of nuclear fusion probability with the temperature of the plasma. The curve that peaks at the lowest energy represents the Maxwellian distribution of energies in the plasma, the curve $\sigma(W)$ represents the probability of fusion with particle energy, and the curve labelled $R(W)$ represents the probability of fusion reactions occurring in a plasma characterized by the Maxwellian distribution shown.

tures can be reached, and the reaction allowed to proceed in a controlled fashion. It has been the task of the plasma physicists in the last twenty five years to try to find out how to do this.

Significant progress has been made. Because high temperatures are used to supply the nuclei with the energy necessary for fusion to start, this effect is called *thermonuclear fusion* and the largest single problem has been that of finding some way of containing the extremely hot gas. It is apparent that this containment will be a great problem, as the melting points of high temperature refractory materials do not exceed the 4150 K of hafnium carbide and tantalum carbide. For this reason alone, the hot gas must not be allowed to come in contact with the walls of the containment vessel. This would also be true for other reasons as well: losses in collisions with the container walls, damage to the container walls in the collisions, cooling of the gas, among other things. The method of containment that seems to have the greatest chance of success is to use the fact that the gas has become a plasma and the electrons and nuclei have separated. Thus we have a fluid which is composed of equal quantities of negatively and positively charged particles, all moving very quickly. These fast moving charged particles can be deflected under the influence of a magnetic field and, at least in principle, careful design of the geometry of the magnets will create a sort of magnetic bottle from which the charged particles cannot escape. This is called *magnet containment.*

The reason why magnetic forces can produce containment of the ionised gas is that they act only on moving charged particles. Also the force is at right angles to both the direction of motion of the particle and to the magnetic field itself. As a result, a charged particle moving at right angles to a fixed magnetic field will move in a circle as shown in Fig. 6.12. A

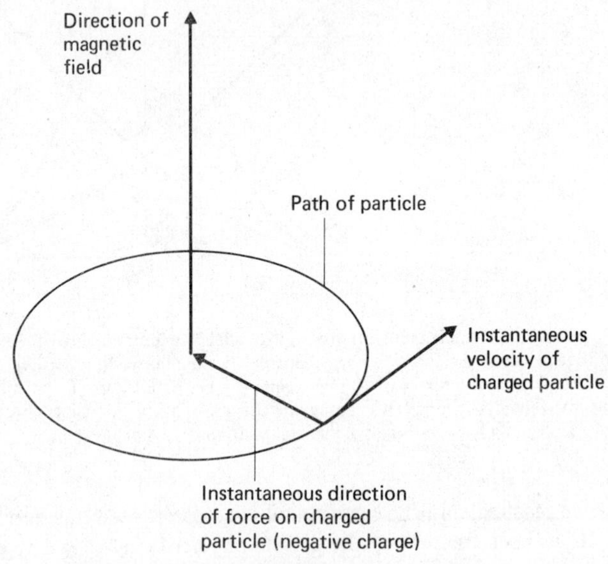

Fig. 6.12 The magnetic force on a moving particle.

magnetic field can either be produced using permanent magnets or by using an electromagnet. The electromagnet consists of windings of wire along which an electric current flows. Associated with the flowing electrons in the wire there is a magnetic field. Increasing the electric current increases this field. By adjusting the geometry of the windings the properties of the magnetic field can be varied. For example, the domestic door chime works by using a cylindrical coil of wire. This is called a solenoid, and when an electric current flows through the wire, a magnetic field is induced inside the cylinder (Fig. 6.13) and forces an iron rod to move and strike the chime. If a solenoid were designed with many extra windings at the two ends, then the effect would be to increase the magnetic intensity and squeeze the lines of force more tightly together at the ends of the solenoid. If this is then used to try to contain the hot plasma, any moving charged particle coming to the edge of the plasma and trying to escape, will find itself deflected and pushed back into the plasma again. This is done by changing only the

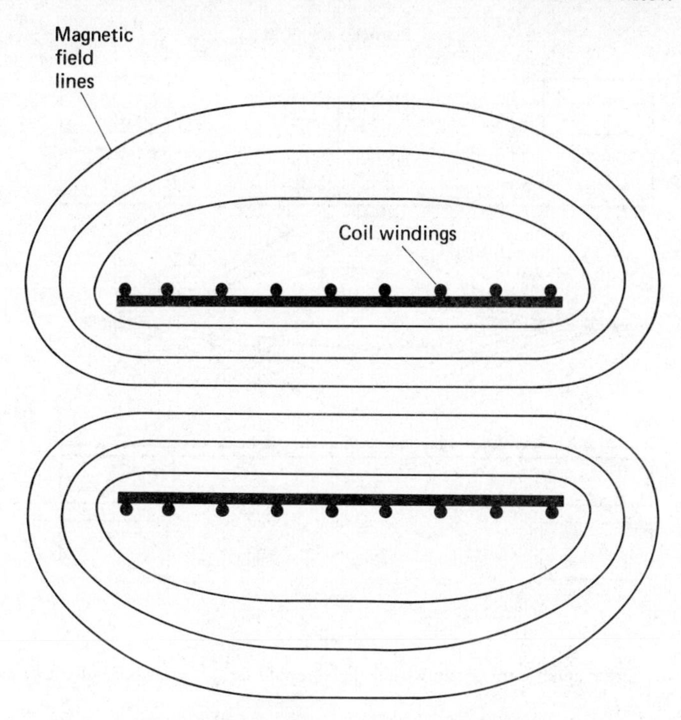

Fig. 6.13 The magnetic field of a straight solenoid.

direction of the particle, and not its speed. Thus, the kinetic energy of the particle and therefore the temperature of the plasma are unaffected. This is called Magnetic Mirror Containment and is shown in Fig. 6.14.

Unfortunately, a bottle of this type has small leaks along the axis of the cylinder at the ends. Any particle starting on the axis of the solenoid and travelling along it will experience no magnetic deflection and will escape. To avoid this, the solenoid is made long and is bent into the form of a doughnut or toroid. Thus there are no ends.

In generating these plasmas, the gas in the bottle is first heated as much as possible by an external means, for example by bombardment by very high energy electrons. If the current through the enclosing coil is now suddenly increased, the magnetic field inside the coil rises sharply. The forces confining the hot ionised gas also increase sharply, and the gas is quickly compressed towards the axis of the magnetic bottle. This is called the pinch effect as the gas is pinched inwards. This sudden decrease in the true volume of the gas increases the pressure, and with it the temperature – everyone knows the way in which a bicycle pump heats up as the air inside it is compressed. It has been found to be possible in this way to reach the temperatures necessary for thermonuclear reactions to take place.

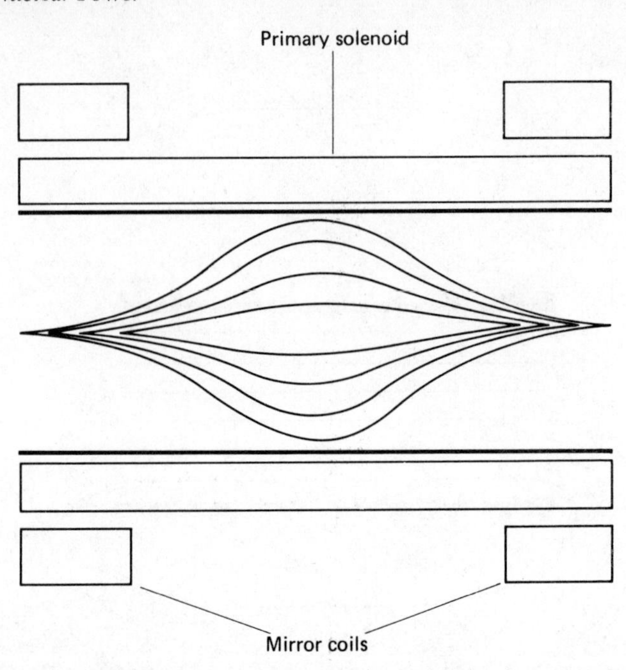

Fig. 6.14 The magnetic mirror in which the solenoid field is enhanced at the ends by extra windings.

The problem that has plagued the whole of controlled thermonuclear research has been the instability of the plasma. Because the gas has been highly compressed to form the plasma, there is a conflict between the outward pressure of the gas and the inward pressure due to the magnetic field. In this conflict, because the fast moving charged particles have their own magnetic fields associated with them, the magnetic field situation becomes very complicated. When this research first started it was felt that the problem would be solved in about five years; this proved to be optimistic. There has been much progress since then and research workers are again confident that the problem can be solved in a few years and that a controlled thermonuclear reaction can be achieved in the laboratory. Even when this is achieved there will still be a long way to go before commercial electric power production will be a reality.

At the moment, the series of machines most likely to achieve the conditions necessary for fusion are the Russian inspired Tokamaks, which are now being built in many laboratories throughout the world. They are toroidal vessels, as shown in Fig. 6.15, and they have the by now conventional toroidal pinch discharge stabilised by a very strong externally applied magnetic field. This magnetic field combines with the poloidal field produced by the discharge current, and produces a resultant helical field. This system is still prone to an instability as the plasma expands radially, and the new development that has helped the performance of the Tokamaks has been the introduction of an outer conducting toroid whose effect is to oppose the growth of these radial instabilities through the effect of eddy currents established in the conducting sheet as the plasma changes shape.

The success of these Tokamaks has been encouraging, and one has reached temperatures of 2×10^8 K for very short periods. This is well above the temperatures required for fusion of deuterium and tritium, but unfortunately, other criteria must be satisfied at the same time.

The progress towards the successful production of controlled fusion is measured by a condition called the Lawson criterion. This states that, in order for the power delivered from the plasma by thermonuclear fusion to exceed the power supplied to it to raise the gas to the necessary temperature, the product of the density of the plasma (in particles per cm^3) and the containment time for the plasma (in seconds) must exceed 10^{14}, the plasma being at the temperature of 100 million degrees centigrade. The actual value obtained for this parameter has risen from 10^8 in the mid-sixties to something over 10^{13} today. The last three orders of magnitude were achieved using the Russian Tokamak toroidal reactor. There is a strong feeling among

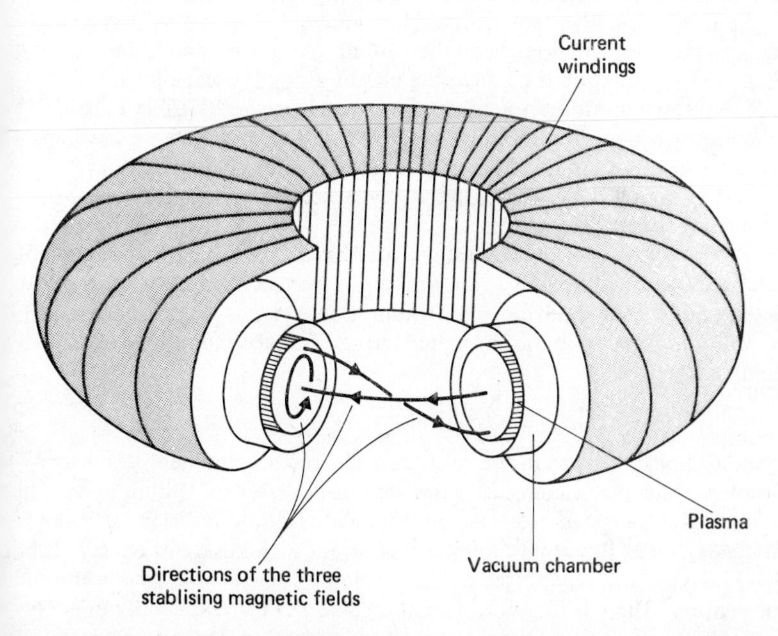

Fig. 6.15 The toroidal solenoid which seems to provide the greatest likelihood of a successful fusion reactor. The structure also includes a large transformer through the centre of the toroid to induce one of the magnetic fields.

fusion research workers that the toroidal reactor will provide the eventual basis for the commercial machine.

There is another possibility being advanced for fusion reactions to be induced in solid samples of fuel, by carefully controlled bursts of laser light – the *inertial confinement* system. This became interesting after computer calculations showed that it would be possible to heat a solid sample to the necessary temperatures in the short time needed so that the pellet of fuel would not explode before the temperature was reached, if a carefully programmed and shaped pulse of laser light was shone on it. This has not yet been achieved in practice, though some reports indicate that thermonuclear reactions have been detected in laser heated samples. Even if this system is successful on a laboratory scale, it is still true that considerable development work on lasers will be necessary before commercial applications could even be contemplated.

Of the possible reactions suggested earlier, the one that will be used first, providing that fusion proves to be possible at all, will be the deuterium–tritium one. This not only releases one of the largest amounts of energy per reaction, but also it can occur at temperatures considerably lower than are possible for the deuterium–deuterium ones. This decreases the containment problem somewhat, and also reduces the losses that are inevitable in any fusion system. The dominant ones are the so-called Bremmstrahlung losses in which highly charged plasma ions emit radiation called Bremmstrahlung, or braking radiation, when they are slowed down in collisions in the plasma. This radiation is a loss to the plasma as it can escape from it, but it will be trapped in the surrounding structure. However, its effect must be taken into account in designing a possible fusion reactor. Bremmstrahlung becomes more and more important as the temperature of the plasma increases, so that the lower temperatures associated with the deuterium–tritium reaction are attractive.

The usefulness of the deuterium–tritium reaction as a realistic source of electric power, assuming that the fusion process can be made to work, is determined by the availability of tritium. Deuterium is a natural component of water, but tritium has to be manufactured from lithium by neutron bombardment

$$n + Li^6 \rightarrow T + He^4 + 4.8\text{MeV}.$$

Is there enough lithium in the world to make the development of nuclear fusion worthwhile? Geological estimates put the reserves of lithium ores, in terms of the energy to be extracted through the fusion reaction, as 1000Q where $1Q = 10^{18}$Btu $\simeq 10^{18}$ kilojoules. Current world usage of energy is abou 0.1Q per year and is expected to rise to about 0.5Q per year by the end of the century. There is therefore something like a 2000 year supply of energy available from the deuterium–tritium fusion reaction. Definitely worth working for! If the process can be made to work and if it can be extended to use just deuterium, the supply of energy is virtually limitless at 10^{10}Q, or something like 50 billion years supply of energy. It does not seem likely that man could possibly use all of that.

Assuming that a fusion reactor can be built, what will it look like? Without even giving detailed models of the construction of the machine, it is possible to predict a few essential features. Firstly the fusion reaction produces not only helium, but also energetic neutrons. This is convenient as long as we are using the deuterium–tritium reaction because, following the philosophy of the breeder fission reactor, we can surround the plasma with a blanket of lithium so that the neutrons escaping from the plasma (which they are bound to do as the magnetic field has no effect on electrically neutral particles) will produce the necessary tritium from the lithium. Also it has been calculated that 80% of the energy of the fusion reaction will be carried by these neutrons so that a special blanket will be needed to convert this neutron kinetic energy into thermal energy, and also to provide a biological shield. Electricity will then be produced by the classical thermal cycle. A few simple calculations give size estimates. It is probable that the magnetic fields will be produced by superconducting magnets. If the temperature of the superconductors is not to be raised by neutron impact and gamma radiation, then a radiation shield some 60–70 cm thick will be needed between the primary neutron attenuator and the magnet. The magnet winding itself, together with its support structure, thermal insulation, etc., will probably be about 50 cm thick. The attenuator will have to be at least 50 cm thick to slow down the neutrons and to breed more tritium. Allowing for a complex vacuum system and adding these sizes together produces a dimension of about 2 m for the structure surrounding the plasma. This is a fairly sizeable installation and so it might as well contain a fairly substantial volume of plasma. Simple estimates based on the toroidal system show that a fusion reactor of this size would produce something over 4000 MW of heat power. This is about the size of the largest electricity generating stations in use today.

The promise of fusion is for cheap, virtually unlimited, and clean electricity. If this were possible then almost every product could be manufactured from basic materials. For example, given sufficient power, it is perfectly feasible to start what would be a present day petrochemical plant from say carbon dioxide and hydrogen, building up the complex organic molecules from these primary constituents. Today this is not economic. It takes too much power to break up the carbon dioxide, and anyway the products derived from the refining of crude oil provide much better starting points. It is only when electricity becomes cheap and oil and coal become either scarce or unavailable that this becomes worth considering.

Radiological Effects of Fusion

The long term hope is that fusion will produce clean electricity. Even so, fusion reactors will entail the handling and control of significant quantities of radioactive materials. However, the possible harmful effects of radiation from fusion power are expected to be less severe than those from fission power.

Only the deuterium–tritium reaction need be discussed in any detail because it is the most probable source of power, and because it produces the widest range of radioactive materials. This inventory starts with tritium itself, which is a beta-emitter with a half-life of 12.3 years. Other radioactivity results from the neutrons produced in the fusion reactions. These will irradiate the entire structure of the fusion reactor and will induce nuclear reactions in the structural materials. Some of these reactions will result in radioactive isotopes being produced. The principal source of the neutron induced radioactivity will come from reactions in the structural material supporting the lithium blanket. This has been proposed as being of either niobium or vanadium. If niobium is used, the principal radioisotope will be an isotope of niobium itself, Nb^{95}. In the vanadium structure the contaminant will be scandium, Sc^{48}. The long lived isotopes will be Nb^{93}(19.6y) and Nb^{94}(2.9×10^4y) for the niobium structure, but for the vanadium structure, only small amounts of various materials will arise through impurities in the vanadium.

As a biological hazard it would seem that the use of niobium reduces the storage hazard from the accumulated waste by about a factor of fifty compared with that from fission waste. It therefore does pose a significant long term hazard. The use of vanadium on the other hand dramatically reduces this hazard as it produces no long lived radioactive isotopes itself under neutron irradiation. In fact niobium is one of the contaminants of vanadium with a concentration of 100 to 1000 parts per million and so the biological hazard can be reduced by three to four orders of magnitude compared to that from niobium. The disadvantage of vanadium is that it would be limited to somewhat lower operating temperatures than niobium which would lead to a less efficient power production cycle. This seems to be a small price to pay for the increased biological security.

7 Inefficiencies

Power production is a major industry in all the developed countries. As consumers we are all accustomed to readily available and reasonably cheap electrical power and refined fuels, and to the benefits which they bring to our homes and factories. It would appear, however, that it is only in times of emergency, for example the 1974 Arab oil embargo, that we give any serious thought to economising on the use of fuel and eliminating wasteful practices. It is precisely this aspect of energy consumption that we shall be investigating in this chapter.

Waste energy has many connotations. The inherent losses in power production have been amply dealt with in earlier chapters but now we shall study how the losses can be recouped and how efficiently the power produced is actually put to work. Similarly, our society produces many waste products, both refuse and sewage, and we shall examine how the wasted energy represented by these products can be recovered, at least in part. Perhaps the greatest energy savings can be made simply by improving the insulation standards of factories, offices and houses, a comparatively neglected topic until recently. And, as a final example, the ubiquitous automobile is at last being designed with economy in mind as well as performance. We shall begin by suggesting some uses for the waste heat produced by power stations generating electricity.

Waste Heat

The conversion of thermal energy to mechanical energy is a most inefficient process. The theoretical limit to the efficiency is $(T_h - T_c)/T_h$ where T_h is the inlet temperature and T_c is the exhaust temperature, both in degrees Kelvin. In practice the efficiencies achieved are much lower than this ideal value and, as we saw in Chapter 5, even the most up to date oil-fired power stations reach only 40%. On the other hand, many of the smaller and older stations are as little as 10% efficient. One way of improving the overall thermal efficiency of a power station is to use the cooling water in district heating schemes.

The term *district heating* refers to the technique of heating a whole section of a city from a central boiler or heat source. The first such schemes to be introduced in the United Kingdom were at Gorton and Blackley in Manchester in 1919, but they were not successful due to pipeline corrosion. Since then various systems have been introduced, the best known of which

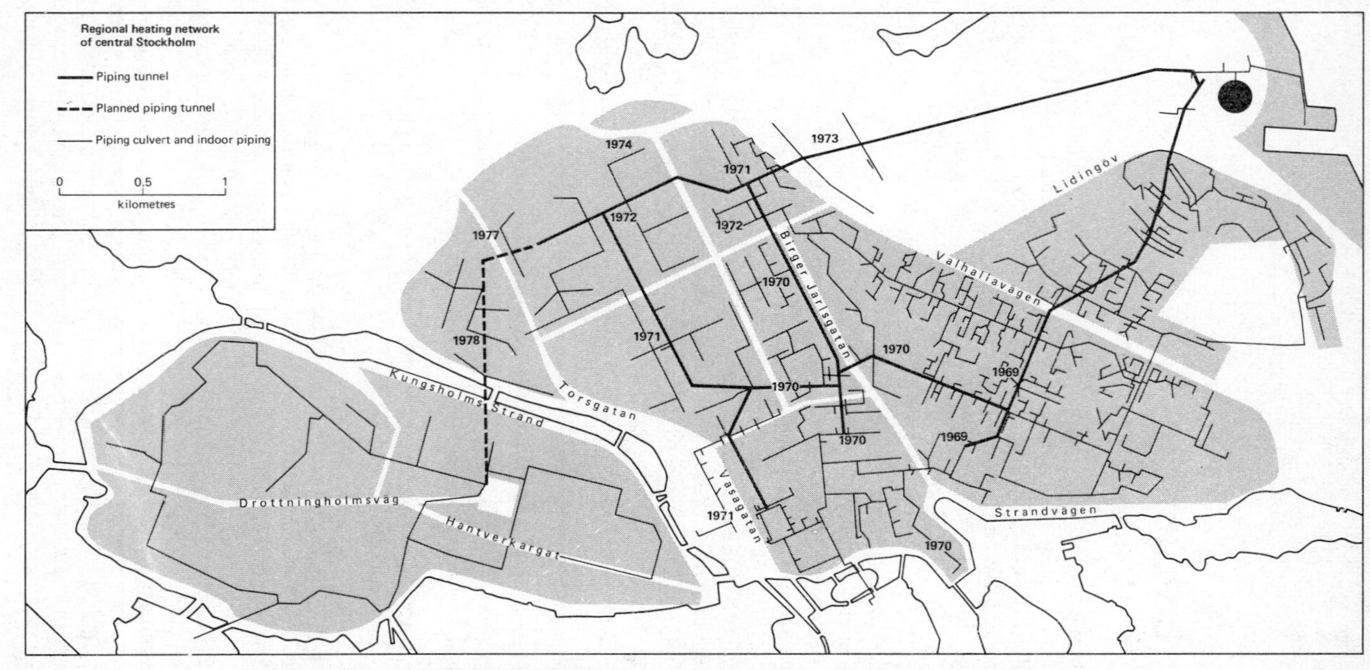

Fig. 7.1 A plan of the district heating network in central Stockholm. (Reproduced by permission of the Stockholm Central Board of Administration.)

is probably the Pimlico scheme involving Battersea power station, London. The heat is taken from the backpressure turbines of the station and hot water is pumped underneath the River Thames through twin pipelines to supply (when completed in 1956) a total of 2403 apartments spread over an area of 31 acres. Nevertheless, district heating has never really become an established procedure in Britain. This is in marked contrast to the United States and many European countries, and district heating has proved to be extremely successful in the USSR and Eastern block countries because of the compact development of housing estates which results from centralised planning.

In the United States steam is normally supplied instead of hot water as the district 'heating' operates air-conditioning plants during the hot summer months. Indeed a recent development there is the supply of refrigerated water for air conditioning purposes via district cooling circuits! The largest district heating system in America is, without doubt, the Consolidated Edison System of New York. Each year this supplies more than ten billion kilograms of steam for space heating, air-conditioning and manufacturing processes in the New York area. The steam is obtained from the back-pressure turbines of five electricity generation stations and from various boiler plants which are usually only required at peak demand during the winter months. The steam is sent out on a distribution grid of underground pipelines with a total length of more than 70 miles.

Perhaps the most dramatic examples of the use of district heating are to be found in the Scandinavian countries. Stockholm for example, has been using this approach since 1903, but major developments have only taken place since 1959. Nuclear power has been used for combined electricity and heat production since 1963. The extent of the Stockholm district heating network is illustrated in Fig. 7.1.

Condensate cooling water is usually released from power stations at a temperature of 30°C, which is low for town heating purposes. If this temperature were raised to 90°C then the use of waste heat for space heating purposes becomes more realistic, but, if we recall our formula $(T_h - T_c)/T_h$, it is clear that an increase in T_c will impair the efficiency. For example, with a T_h of 600°C the maximum theoretical efficiency drops from 65% to 58%. Consequently, we must strike a balance between how much electricity and how much space heating we wish to have.

The preferred method of producing heat and electricity simultaneously in power stations is to use the Intermediate Take-Off Condensing (ITOC) turbine whose layout is depicted in Fig. 7.2. Essentially the turbine is divided into two parts and steam can be bled from the ends of the individual parts or at a number of intermediate points on the turbine units. When the heating requirement is large, steam is taken off at the high pressure end and is either used directly for heating or indirectly by heat exchange with the water of a district heating system. When the demand for heating falls, the by-pass steam is reduced and the ITOC turbine behaves like a normal con-

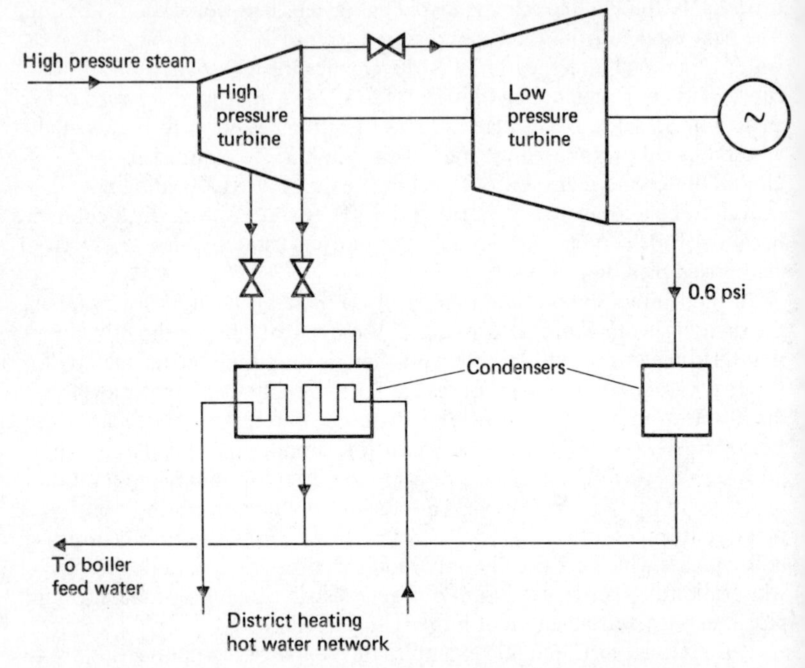

Fig. 7.2 A schematic representation of an ITOC (Intermediate Take-Off Condensing) turbine operating a district heating scheme.

densing turbine. During periods of peak electricity demand the by-pass steam lines are shut and the turbine is allowed to run on a completely condensing cycle, so generating as much electrical power as possible.

It is worth noting in passing that district heating is by no means restricted to systems based on waste heat from power stations. A centralised boiler unit can be designed to cope with a lower grade of fuel than can a single domestic system. Moreover, such a centralised boiler can also handle unconventional fuel such as refuse and sewage (see p. 151). The siting of a single boiler unit for a town or region of a town can be selected so as to minimise transportation costs for the fuel concerned. Against this, district heating has the disadvantage of requiring an expensive pipeline network. District heating schemes will be most economic when the networks are large, when there is high density housing, when the system is being installed in a new housing development, and when a high percentage of buildings within the network area are actually connected to the network.

The diesel engine and the gas turbine also offer possibilities for small-scale combined heating and electricity generation schemes, the so called 'total energy concept'. The diesel engine can burn both heavy and light oils with an efficiency of up to 40%. Some 60% of the total waste heat can be recovered from the high temperature exhaust gases and from the cooling water, which gives a total efficiency of 75%. The gas turbine is only 15% efficient for electricity generation but, assuming the same figure of 60% for the fraction of waste heat recoverable, we still end up with an overall efficiency of close to 70%. If electrical power is more important than the heat produced then the diesel engine is the best bet and vice versa. Using this total energy scheme it ought to be possible for a factory, or group of factories, to become independent of the electricity supply. The main drawback to this, at any rate in a country such as the United Kingdom, is the large premium charged by the authorities for standby electricity. Thus to allow for repairs or routine servicing of the equipment it will be necessary to instal much more capacity than is actually required.

Waste Products

In modern industrialised societies the problem of dealing with refuse and sewage is becoming increasingly embarrassing for the authorities concerned, especially in densely populated urban areas. In the United Kingdom it is estimated that 20 million tonnes of refuse is produced each year. To spread this refuse evenly on rubbish tips to a thickness of a metre or so would mean submerging many thousands of acres of land beneath this none too pleasant material. Admittedly poor quality land has been used in the past but this is becoming more and more scarce. In any case, much more than just the area of the tips is blighted: who wants to live beside a mound of rubbish? One way of dealing with the problem is to decompose some of the rubbish into manure but this is expensive and slow. The solution of the future may well be to burn the refuse.

Refuse from countries such as the USA and UK has a calorific value of about one-third that of coal. Incineration reduces the refuse to between 10 and 15% of its original volume and of course renders it completely sterile. Moreover, this stable residue can be used as hardcore in the building and civil engineering industries. However, it would seem that small incinerators dealing with less than 10 000 tonnes of rubbish per annum are not economic. It has been estimated that the cost of disposing of rubbish in this way exceeds the cost of controlled tipping by 50%. The operation of large incinerators, on the other hand, is far more favourable given that the heat generated can be used in district heating schemes.

The burning of refuse in incinerators poses some unique problems. The difficulties arise from the dampness of the material (up to 40% water content), its low but extremely variable density (averaging 300 kg m^{-3}) and the fact that the refuse is a mixture of widely varying components ranging from dust to old springs, from plastic to wood and even the ubiquitous old boots! Refuse incineration plants must therefore be designed in such a way as to sort the material into non-combustible and 'fuel' components, dry the fuel thoroughly before combustion, and then agitate it so as to ensure complete combustion. In recent years fluidised combustion of a ground coal–refuse mixture has also been developed. Most incineration plants need to use some auxiliary fuel whenever the refuse has a lower than usual calorific value and sump oil from service stations is commonly used for this purpose.

Although local authorities in the UK are only just beginning to take an interest in refuse incineration, both for refuse disposal and as a way of collecting useful energy, it is already an established practice in Europe. Denmark, for example, has more than 400 district heating systems of which at least 40 are refuse incinerators with attached boilers. In Paris the refuse incinerators are solely responsible for supplying the municipal heating system between March and October; only in the winter months do the conventional power stations have to be used for this purpose. In London some electricity is generated from refuse, at the Greater London Council's Edmonton incinerator. This burns the refuse in high temperature boilers and uses the steam produced to generate electrical power.

It must be admitted that refuse incinerators are expensive to build, maintain and operate. They are also very inefficient; typically heat losses in the flue gases and in the slag amount to 40% of the heat generated. Consequently, the sale of heat from these plants does not pay for their running costs. In fact the local authorities whose job it is to dispose of the rubbish invariably have to subsidise the operation of their incinerators. Nevertheless, it must be worthwhile for a large city to get some energy back from its refuse, especially as this method reduces the number of unsightly tips into the bargain.

Recycling is a partial alternative to incineration of waste materials. Many waste products which have now reached the end of their useful lives were originally produced by industrial processes involving high energy consumption. Thus the rotting hulk of an abandoned automobile represents a waste of iron and a waste of the energy that was used in turning the iron ore into steel. By recycling these products, if only to a point just beyond the raw material stage in the manufacturing process, a considerable energy saving can be made. The incentive for recycling is of course economic and the fact that 40% of paper, 50% of crude steel and 70% of lead used in the United Kingdom is at present recycled simply indicates that the recovery of such materials is economically extremely attractive. Furthermore there is the secondary benefit of resource conservation. Many other recovery processes may well become economic as the availability of world resources changes. Changes in fuel costs may also encourage recycling. Aluminium production provides a good example. Aluminium prepared by electrolysis from raw materials 'costs' 17 000 kWh per tonne whereas recycled aluminium can be purified for 450 kWh per tonne.

There are two major problems associated with recycling. The first is the separation of the various components of rubbish. One technique which currently appears to have considerable promise is fluidised bed separation.

A bed material of, say, iron powder lies in a trough which is filled with rubbish and then vibrated. As a result of the vibration, the high density wastes are encouraged to fall through the bed material while the low density wastes remain floating on the surface. ('High' and 'low' are of course relative to the density of the bed material.) It is already feasible to separate metal and plastic mixtures with a throughput of one tonne per hour using this technique. The second problem is more correctly a misconception. It is not accurate to say that all of our waste products could be recycled given sufficient economic incentive. For example, although a high proportion of paper is already recycled, it is not possible, for technical reasons, to recycle good quality paper as such. Each successive manufacturing process degrades the quality of the paper until it can only be used for cardboard manufacture. Similarly, problems arise in glass manufacture which make it impossible to recycle more than about 25% of production (of course, bottles can be reused!).

The Heat Pump

Space heating accounts for a sizeable fraction of energy consumption in countries such as the United Kingdom and the United States. In the latter case some 18% of total energy consumption is allocated to this need. Clearly there is considerable scope for conservation measures in this area, and, in particular, a somewhat conservative estimate by OECD (1974) indicates that 10% of all consumption in this sector could be eliminated by the use of heat pumps.

We are all familiar with the refrigerator as a means of keeping food fresh in our homes and shops. In this machine heat is transported from a cool environment, the inside of the refrigerator, to a warmer environment outside. At first sight this seems to contradict the natural tendency for heat to flow from a hot to a cold body, but, as we shall see, energy is consumed in driving the machine and 'pumping' the heat against the temperature gradient.

A refrigerator is used to extract heat from an enclosure. When the procedure is reversed so as to warm the enclosure by extracting heat from its surroundings, we call the machine a 'heat pump'. This is by no means a new invention, an impression that might be gained from the limited use of heat pumps today. The principle of warming a house by refrigerating the atmosphere outside was enunciated by Lord Kelvin as long ago as 1852, but 75 years were to elapse before the first details were published of a practical heat pump which was built by Haldane. He used it to heat his house in Scotland by cooling the outside air and the local water supply. Perhaps it should not surprise us that the Scots, with their renowned financial acumen, were the first to recognise the economic advantages of the heat pump!

Most refrigerators or heat pumps use a vapour compression cycle of the type depicted in Fig. 7.3. The refrigerant is allowed to pass via a throttling

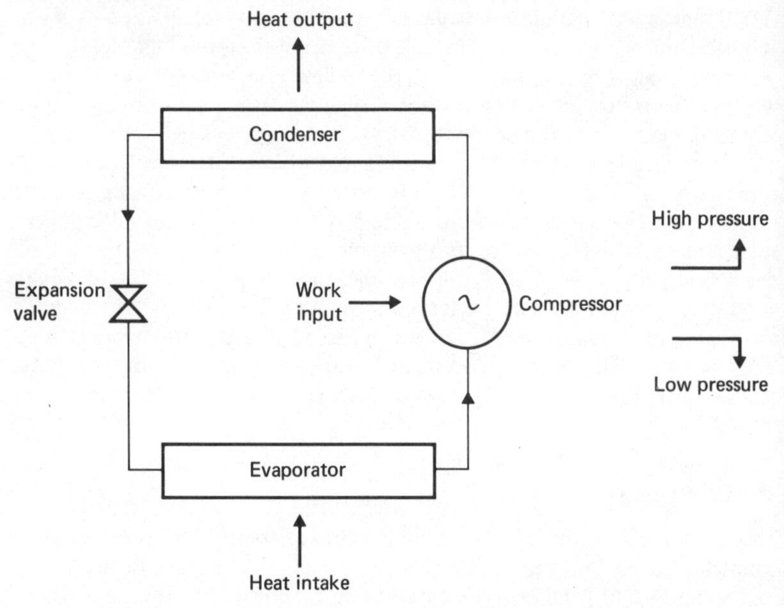

Fig. 7.3 The vapour compression cycle for a refrigerator or heat pump.

valve to an evaporator after which it is compressed and returned to the condenser where it loses heat and is thus returned to its original state. In the condenser the pressure and temperature are high, and the refrigerant gives up its heat by condensing to a liquid, releasing the latent heat of vaporisation. Expansion of this hot liquid through the throttling valve lowers its pressure and temperature and in practice partially vaporises it. This vaporisation is completed in the evaporator, where the refrigerant extracts its latent heat of vaporisation from the materials to be cooled. To complete the cycle the vapour is now compressed – and this is where an external power supply is required – and its pressure and temperature rise once more. This vapour is cooled and liquefied in the condenser and the heat given up to the warm environment. Thus heat is transported from the cool to the warm region using the mechanical energy from the compressor. It should be noted that at no time does heat actually flow directly against a temperature gradient. The appearance of a practical heat pump may be gauged from Fig. 7.4 which shows a packaged air-to-water device. This means that heat is extracted from the atmosphere and delivered into hot water which may then be circulated through a dwelling's radiator system.

The performance of a heat pump may be specified by the heating effect produced in a certain enclosure for a given input of mechanical energy. The *coefficient of performance,* CP, for a heat pump is defined as the ratio of the heat delivered to the net work done in each cycle. As energy must be conserved overall the net work done is the difference between this heat delivered and the heat extracted from the cool environment. The theoretical limit for CP is $T_h/(T_h - T_c)$ where T_h and T_c are the temperatures between which the heat pump is operating. For example, if heat is absorbed from river water at 2°C (275 K) and used to warm a room to 22°C (295 K), then the theoretical CP is 295/20, or nearly 15! Of course any real heat pump

Fig. 7.4 A prototype commercial heat pump under test in the authors' laboratory. (Reproduced by permission of Steel Link Engineering Co. Ltd.)

will not achieve this maximum level of performance. There will be inefficiencies in the compressor itself and also in the heat exchangers used to transfer the heat between the refrigerant and the surroundings. In addition, practical systems require pumps and fans (to promote water flow through or air flow across the heat exchanges) thus increasing the energy consumption of the machine. In practice, coefficients of performance of between 2 and 7 have been achieved by commercial installations.

The heat pump is at its most efficient when the temperatures of the heat source and heat sink differ only slightly. For this reason it is preferable to use heat pumps in conjunction with low temperature heating systems. The ideal low temperature heat source would have a constant temperature throughout the 'heating season' that is almost as high as the target internal temperature. It goes without saying that such sources are hard to come by and so we must reconcile ourselves to using less suitable ones. Water from rivers or lakes is one possibility provided it is not so cold as to freeze on the heat transfer surfaces of the evaporators. Thus the heat pump is ideally suited to working in a climate like that of the British Isles. Air can also be used as a low temperature source, but its temperature can vary considerably even during the course of one day. This means that the heat pump would have to be designed with sufficient capacity to cope in cold weather, but for the greater part of the heating season it would be running below its full output. To avoid this, auxiliary supplies of heat could be used to assist the heat pump when demand is at a maximum.

Suitable refrigerants for use in heat pumps should have boiling points in the range −40°C to −10°C. Ammonia, with a boiling point of −33°C, was commonly used in refrigerators but had the disadvantages of being inflammable, toxic and corrosive. To avoid these problems the fluorocarbon refrigerants, known by the trade names Freon or Arcton, were introduced. The most common ones in use today are Freon 12 (CCl_2F_2, bp −30°C) and Freon 22 ($CHClF_2$, bp −41°C).

By using heat pumps to warm buildings we are effectively increasing the efficiency of electricity production. The average efficiency of electricity generation is something like 30%. However, if we use this electricity to power the compressor of a heat pump with a CP of say 3.3, we can obtain 3.3 times as much heating as would be the case if the electricity were used in conventional appliances. This means that this particular heat pump can generate the same amount of heat in the building concerned as was generated by the combustion of the fuel in the power station. In addition, the waste heat generated by the power station could, as we have suggested above, be used in a district heating scheme. Consequently, it is possible to obtain an amount of heat from the fuel which is greater than the calorific value of the fuel – an 'efficiency' of more than 100%! Even better results can be obtained if an internal combustion engine drives the compressor and the waste heat from the exhaust gases is used for space heating. For example, with a 40% efficient diesel engine linked to the same heat pump, the amount of heat delivered is 132% of the heat content of the diesel oil. The exhaust gases can be passed through a heat exchanger with, say, 60% conversion efficiency to give a further 36% (0.6 x 60) of the fuel's heat content towards space heating. Our grand total of heat delivered is therefore just under 170% of the energy content of the diesel oil!

With such good reasons for installing heat pumps, why is it that they have not been more widely used? Capital cost of the installation is one reason, but this will become less prohibitive as fuel costs rise. However, one has the feeling that the main reason is an unwillingness, even on the part of some scientists, to believe that the heat pump really works, that we can get 'something for nothing'. This is, of course, precisely what we are not getting. We are simply using a small amount of energy to drive a larger equivalent amount of heat against a temperature gradient. It is odd that people are so ready to accept that a refrigerator can work, but not a heat pump. (After all, we could use our refrigerators as heat pumps simply by fitting them into a window frame with the refrigerator door open to the outside!) This is not to imply that the heat pump has been overlooked by heating engineers. Indeed the current United States market for domestic units is estimated at 150 000 per annum.

Heat pumps can be used even more efficiently if the heat extraction part of the process can be put to good use. For example, refrigeration and water heating can be accomplished in the same unit by putting the evaporator of the heat pump into the space to be cooled and immersing the condenser in the hot water storage tank. Similarly, by a suitable system of pipes and ducts the heat pump can be used as a space heater in winter *and* as an air-conditioner in summer.

If this section is thought by the reader to be strongly biased towards the use of heat pumps, then it must be admitted by the authors that he is quite correct. Indeed it can be argued that figures for the efficiency of space heating based on electrical power should never be quoted without a 3 : 1 improvement on the assumption that a heat pump will be used. In addition, the heat pump reduces the amount of heat liberated into our environment. Instead of generating say 6 kW of heat within a typical house we can simply transfer 4 kW into the house from outside by using 2 kW of electrical power. Thus only 2 kW of heat is added to the local environment; the loss of 4 kW is made up when the heat is returned to the environment through the walls of the house. Similarly heat pumps which use the coolant water or effluent from industrial plant as the low temperature heat source are the only means by which this low-grade heat can be used and, perhaps more important, allow us to return this water to the environment at something like its original temperature.

Insulation

It is well appreciated by governments as well as technologists that the simplest way of reducing the consumption of fuel for space heating is to improve the thermal insulation of our buildings. Spurred on by the recent oil crisis the British government's Department of the Environment proposed in January 1974 that mandatory requirements for stricter standards of insulation be introduced in the building trade. The Department suggested that the maximum permitted heat loss through the walls of a house be reduced from 1.7 to 1.0 W m^{-2} $°C^{-1}$, and that heat losses through the roof be limited to only 0.6 W m^{-2} $°C^{-1}$. Of course this would only apply to new houses but at any rate the proposed regulations, although a distinct improvement on existing standards, would still allow a considerable quantity of valuable energy to escape into the environment. By the middle of 1976 these proposals had not been implemented!

Before discussing insulation techniques we should perhaps give some thought to the question of just how warm we want our houses to be. During the 1974 energy crisis we were urged to reduce the thermostat settings in our houses and offices. It came as something of a surprise when many of us discovered that we could survive such apparent hardship without experiencing too much discomfort. In fact many central heating systems are normally run at excessively high temperatures, causing the air to be very dry and uncomfortable (unless humidifier units are installed) so it would seem that we can all make savings simply by settling for a reduction in room temperature.

Similarly, many savings can be made, both in heating and overall electricity consumption, by a conscious effort on the part of the individual to use appliances efficiently, and to avoid senseless waste. For example, the housewife can play her part by using pressure cookers when possible, by keeping lids on saucepans and by using the correct size of cooker ring for each saucepan. Ensuring that light bulbs, light shades and windows are free from dust and dirt will contribute towards more efficient lighting. Windows should only be opened for ventilation and not for cooling an overheated room – it is much better to control the heating correctly in the first place. In industry it may be worthwhile to install photocell lighting-control, so ensuring that lights are not left on during daylight hours. As a further refinement on this theme, it is prudent to switch lights near windows separately from those in the interior of a building. Again, the installation of fluorescent light fittings can be a significant economy as a 40 W fluorescent tube is equivalent in lighting capability to a 150 W tungsten filament bulb (although it costs more energy to manufacture). It would be an easy undertaking to extend this list almost ad infinitum, and most electricity utilities now publish booklets on energy savings, but, while it is true that these are all practical and sensible economy measures which we as individuals can adopt, much more significant savings can be made by improving the insulation standards of buildings and by applying radically new design concepts for heating and lighting requirements.

Until recently, most of the effort in house insulation was expended in reducing heat losses through the roof. Most new houses are adequately insulated in this respect and it is a fairly simple matter to improve conditions in older property. One standard technique is to fill the spaces between the joists on the floor of the roof space with polystyrene chippings (which have low thermal conductivity) and to cover these with a layer of glass fibre insulation. The top of the roof space is usually lined with roofing felt which, in addition to its insulation behaviour, provides some weatherproofing beneath the roof tiles. Needless to say, techniques vary with the type of house and the method which we have described is particularly prevalent in the United Kingdom where most houses are built with bricks and tiles. In wooden frame houses the basic material is itself of low thermal conductivity, so long as it is treated to resist damp.

Heat losses through walls and windows can be enormous, as is illustrated by the calculated data of Table 7.1. Glass is the worst offender, but heat losses through windows can be halved by using double-glazing, because of the extremely low thermal conductivity of the trapped layer of air

Table 7.1 Heat losses through walls and windows

Calculations are based on an internal temperature of 20°C and an external temperature of 0°C

Structure	Heat Loss (W m^{-2})		
	0	parallel wind speed	30 mph
(a) single-leaf brick wall, plaster on inside*	70		115
(b) double-leaf brick wall, plaster on inside*	44		60
(c) cavity brick wall (5 cm cavity), plaster on inside	28		36
(d) as in (c) but with urea formaldehyde foam in the cavity	10		11
(e) single pane 0.3 cm glass	80		150
(f) double glazing with two 0.3 cm panes separated by a 2.5 cm air gap	45		65

*Standard British brick of 4½" (11.3 cm) with ½" (1.3 cm) plaster.

(0.03 W m^{-1}°C^{-1}) or vacuum. Similarly, it pays to build two brick walls, one just inside the other, and so capitalise on the insulating properties of the air in the cavity. (An additional advantage, which actually was the original purpose of this mode of construction, is that it checks the spread of damp through the wall.) The figures quoted in the table are only approximate values and fail to take into account such factors as the moisture content of the bricks. In wet weather this will be high but will be reduced when the tangential wind velocity is also high. Thus the data given should only be treated as a guide to the various heat losses.

At first sight it would appear that most gains could be achieved by installing double-glazed windows, but although cavity walls are better insulators than single-pane windows, there is usually a greater wall area than window area in a given house. Consequently, more heat is lost through the walls than through the windows and improvements in wall insulation may pay much greater dividends than double-glazing. It is now becoming increasingly popular to insulate houses by filling the wall cavities with insulating material such as urea formaldehyde foam. (The insulation is still provided by air, in this case trapped as bubbles in the foam when it sets, but the urea formaldehyde is itself a good insulator.) The advantage over an ordinary cavity wall is that draughts in the wall space are eliminated and heat transfer can only occur by conduction and not by convection. Naturally it is cheaper to insert the foam into the cavity while the house is actually being built, but many firms now specialise in cavity wall insulation of established property. The foam is simply pumped through a number of holes drilled in the outside wall. One objection to this technique of cavity wall insulation is that the foam may tend to absorb water so that damp may penetrate to the inner wall when there is heavy rain. However, as long as a closed cell foam is used there should be no question of moisture penetration assuming that the cavity has been filled by competent contractors.

With the walls properly insulated we can now turn our attention to the windows. We can reduce the heat loss through windows by decreasing the window area of the house but this is not necessarily a good thing to do. In south-facing rooms windows produce a greenhouse effect during the day and this usually leads to a net gain during a 24 hour period. Apart from installing double-glazing, heat losses at night can be reduced by drawing heavy curtains across the windows or, to be slightly old-fashioned, by closing wooden shutters in front of them.

The expense of cavity wall insulation or double-glazing is to some extent wasted if the house leaks, that is if it is prone to draughts. Fortunately, this is a fairly simple problem to correct even if it does involve tedious work. Window and door frames can be lined with a material such as sponge rubber to ensure an airtight fit. Similarly, the flue of an open fire can be blocked off when not in use.

To illustrate the magnitudes of the heat losses from a typical house, Fig. 7.5 shows schematically the losses from an average size bungalow before and after the adoption of the insulation practices described in this chapter. As a final point, it is worth noting that the economic order of priority for insulation is currently draught-proofing first, then roof and wall insulation, and lastly double-glazing. In fact at present prices double-glazing takes more than ten years to pay for itself.

Once a house has been effectively insulated many unexpected sources of heating begin to play a major role in reducing the domestic fuel bill. For instance, an individual sitting at rest is losing some 200 W of heat to his environment, which means that five people sitting in a room generate as much heat as a 1 kW electric heater! Even the lights and television set begin to be an important source of heat when the insulation is improved sufficiently. In fact, electric lamps are extremely inefficient sources of light; most of the electrical input is converted into heat. The most up to date building designs attempt to use this in the heating supply. This is especially true in large buildings such as office blocks. The interiors of these require continuous artificial lighting, and air-conditioning is necessary to maintain the temperature at acceptably moderate levels in these regions. The alternative is to integrate lighting, heating and ventilation into one scheme. This can be done by incorporating ventilation slots into the light fittings. Air passed through these slots extracts the heat from the interior light fittings

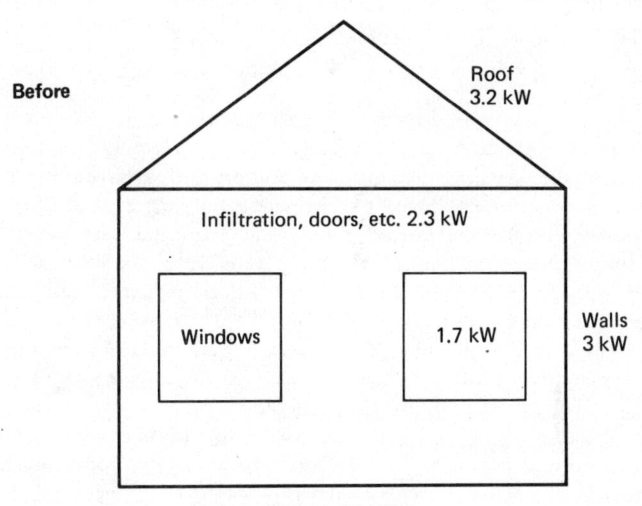

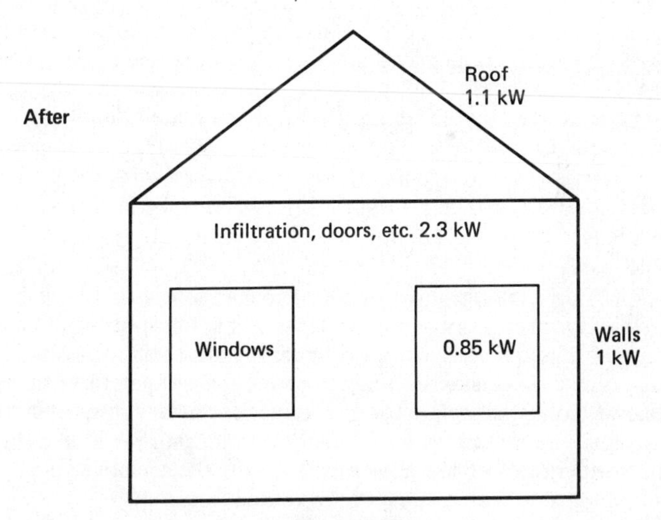

Fig. 7.5 Heat losses from a typical 10 sq. m. bungalow before and after insulation procedures. The particular procedures chosen are 50 mm glass fibre roof insulation, urea formaldehyde foam wall insulation and double glazing.

and distributes it around the whole building. In this way much conventional heating can be avoided and the building can be warmed simply by switching on the lights!

Transport

Since the 1973–4 oil crisis forced several Western governments to impose some form of petrol rationing, numerous economy devices have appeared on the market. Unfortunately, many of these are totally useless but there are some which can reduce fuel consumption. One technique, used in aero engines during the last war, involves mixing some water vapour with the petrol–air mixture and a similar improvement in performance can be noted by any motorist on a damp morning! Alternatively, the fuel can be preheated before it reaches the engine's combustion chamber. By helping to ensure complete combustion of the hydrocarbon fuel this leads to better efficiency and less pollution into the bargain.

It must be admitted though that the most influential people as far as fuel consumption is concerned are the motorist himself and the automobile manufacturer. The driver, as we are all aware, can greatly influence his consumption figures simply by changing his style of driving, by avoiding heavy acceleration and braking. The manufacturer, on the other hand, has it within his power to shift the emphasis from high performance automobiles to vehicles which have economy as their selling point, a trend which is already established. In this he may presumably be influenced by the move towards low blanket speed limits in most countries, a trend that will presumably be reinforced by the arguments of the road safety lobby.

Of course there are many who would say that we should save our precious petroleum reserves by abandoning the private car altogether and encouraging more extensive use of public transport. This is all very well for someone living in a large city such as London or New York. But for those of us who are unfortunate enough (or should one say fortunate) to live in rural areas, deprivation of private transport would mean being forced to drastically alter our life style, for in sparsely populated areas public transport would have to be heavily subsidised if it were to provide an adequate service. A thirty-five seat coach is more expensive to run with driver and one passenger than is a private car with only the driver! The importance of private transport in rural areas is emphasised by the extremely high car ownership per head of population in Northern Ireland and East Anglia despite their relatively low prosperity compared with the rest of the United Kingdom.

Returning to vehicle design, it is apparent that there are in existence three or four simple techniques for improving fuel consumption, all of which lead to additional benefits. Firstly, the use of radial ply tyres, as well as improving handling and tread life, can reduce fuel consumption by 10% overall. Secondly, improvements in load to engine matching and the installation of an overdrive gear can bring about decreases in consumption by 20% overall and 20% in high speed driving respectively. Moreover, they both lead to less engine wear. Thirdly, an aerodynamic body design can reduce consumption by the order of 5% and is often aesthetically pleasing. Unfortunately, there appears to be little hope of a radical improvement in the efficiency (about 20%) of the internal combustion engine, although research is being carried out in the design of stratified charge, Stirling cycle, Rankine cycle, and gas turbine engines. A major development of the future, already mentioned in Chapter 3, is the electric car. When one includes all the energy conversion steps implicit in the operation of an electric car the *overall* efficiency of such a machine is about 15%, compared with the 10% *overall* efficiency of the conventional automobile. This is certainly a large relative improvement but it must be set against the inflexibility and limitations of the electric vehicle.

Taking a slightly more long term and radical view, it is possible that significant savings in the transport sector could be made by increased reliance on telecommunications. It is already commonplace for individuals and even firms to place telephone orders on the basis of catalogue information. With increased acceptance of this method of conducting business, and with improvements in telecommunications technology, it could be that many of the short and relatively expensive automobile journeys made for shopping or for business purposes will be eliminated. Similarly, there is a need for extensive study into overcoming the problems of commuting in large urban conurbations. The standard pattern of city development is now well established. By the time most towns have grown into large cities, industry, business and commerce is concentrated in the centre while the working population lives in the suburbs. Most of the effort in transport policy is directed at more efficient transfer of these suburban residents to and from the city centre. But now business and industry can only expand in that city by moving to the suburbs, so that the planners have the new problem of coping with the suburb to suburb commuter. Surely the answer must lie in long-term urban planning which has the aim of integrating residential and work activities, and which has the spin-off of reducing energy consumption in transportation.

Investment in Energy Saving

Clearly the scope for energy saving is linked to the economic climate. Energy saving measures themselves cost money, which must be invested before any return can be seen. As the price of fuel rises, so the cost of energy saving measures becomes less prohibitive in relative terms. We can illustrate this in the case of savings in the industrial sector.

There are many techniques which we have failed to mention and which could play a valuable part in reducing industry's fuel bill. For example, waste heat recovery ought to be employed in every industry where there are both high temperature and low temperature processes. Improved process technology and better control of industrial processes can both bring about energy savings. As final examples, the use of variable speed instead of fixed speed motors and the introduction (or should it be re-introduction) of hydraulic lifting systems are both apparently trivial modifications, but they can effect considerable energy savings. Thus in the United Kingdom it is estimated that the widespread use of variable speed drives could lead to a 3% saving in *total* energy consumption. Why is it that these ideas are not being introduced more rapidly? The answer, ironically, lies in the renowned cost-consciousness of the industry sector. However trivial or complex the modification, it costs money, and moreover the industrialist may be faced with introducing efficient plant to replace plant which still has a considerable life expectancy. Energy costs often account for only a small fraction of final production costs and the industrialist is constrained by the higher maintenance and operating costs of any new highly efficient equipment he may introduce. It is clear, then, that it is the accountant and not the technologist who decides how much energy a given firm is going to save.

This leads us to a very significant point indeed. Should energy savings be costed financially or in terms of energy? Let us take the example of cavity wall insulation. At current prices the savings which accrue to the householder should pay off his investment within three years (and this has certainly been confirmed by the authors' combined experience). But the urea formaldehyde foam is prepared by an energy intensive process and from petrochemical feedstock to boot! When this and the energy consumed during installation are taken into account, some estimates have it that eight years elapse before the initial energy investment has been paid off. The moral is that we should cost energy in its own 'real' terms by means of energy accounting if we are indeed set on a realistic course of energy saving. (An additional advantage of this approach is that the calculations are inflation proofed. Presumably, in the particular example above, 'inflation' or 'deflation' is caused by a trend towards higher or lower room temperatures respectively.)

What is the likely total financial cost of the measures we have been advocating? The OECD estimate that if 20% of the Western world's energy consumption is to be saved by 1985 a cumulative investment of more than three hundred billion dollars will be required. This is big money indeed, but it must be remembered that such investment will not only save money in the short to medium term, it will also give us time to develop the technologies of the future and thereby safeguard our civilisation.

Appendix

Units

There is an unfortunate amount of confusion in the popular literature between the related concepts of fuel, energy and power. A fuel can be defined as a substance which contains energy in an orderly (low entropy) form and which can be made to release that energy in a usable manner. Power is the rate of use or release of energy. Thus although they are related, they are not the same.

To be strictly correct, there cannot possibly be an energy crisis, because according to the first law of thermodynamics, energy is always conserved; it cannot be created or destroyed, but only converted from one form to another. Properly speaking, what we suffered in the early nineteen seventies was a fuel crisis – or perhaps even more correctly, a fuel price crisis.

A study of the units in which the three concepts are measured may help to clarify our thinking.

Energy is measured, in SI units, in *joules*. A joule is the work done by a force of one newton in moving a body through a distance of one metre in the direction of the force. Other units of energy are shown in Table A1.

Power is measured, in SI units, in *watts*. A watt is one joule per second. Other units are shown in Table A2. Strictly speaking, the nature of the energy released needs to be specified in any definition of power. Sometimes this is obvious, but an example of the sort of situation in which care is needed is in nuclear power stations, where the power station as a whole is rated in terms of installed electrical generating capacity or of maximum electrical power sent out (these two differ slightly because the power station uses some of its own power to run its plant) while nuclear reactors are very often rated in terms of their thermal power output. Since, as we have seen, there is inevitably a loss of efficiency in converting heat into mechanical or electrical energy, the thermal power rating and the electrical power rating can be very different.

Fuel can be measured in units of energy per unit mass, and a reserve of fuel can be measured in units of energy. Strictly, however, the means of release of that energy needs to be specified; for example, deuterium (heavy hydrogen) used as a fuel in a thermonuclear power station has an immense energy per unit mass, while the same deuterium used as a fuel in a domestic gas cooker (it burns just like hydrogen) has a much lower energy per unit mass.

Table A1 Conversions of Units of Energy

Basic unit . . . joules.	
1 calorie	4.2 joules (unit used in chemistry)
1 Calorie (kilocalorie)	4.2×10^3 joules (unit used in diets)
1 Btu (British thermal unit)	1055 joules or 252 calories
1 therm	1×10^5 Btu or 1.055×10^8 joules
1 mtce (million tons of coal equivalent)	2.55×10^{13} Btu or 2.69×10^{16} joules
1 mtoe (million tons of oil equivalent)	4.25×10^{13} Btu or 4.48×10^{16} joules
1 barrel (of oil) (42 US gallons)	5.62×10^6 Btu (approx.) or 5.92×10^9 joules (approx.)
1 watt hour	3600 joules or 3.42 Btu
1 kilowatt hour	3.6×10^6 joules or 3416 Btu
1 MeV (million electron volts)	1.6×10^{-13} joules

Table A2 Conversions of Units of Power

Basic unit . . . watts.	
1 Btu $hour^{-1}$	0.293 watts
1 horsepower	746 watts
1 kilocalorie $hour^{-1}$	1.167 watts

Bibliography

References are listed by chapter and are arranged in the order in which their subject matter is discussed.

General

MCMULLAN, J. T., MORGAN R., MURRAY, R. B., (1976) *Energy Resources and Supply,* Wiley Interscience, London.

ORGANISATION FOR ECONOMIC COOPERATION AND DEVELOPMENT, (1974) *Energy Prospects to 1985,* OECD, Paris.

(1974) *Energy Conservation – a study by the Central Policy Review Staff,* HMSO, London.

(1971) *Energy and Power,* Scientific American book, W. H. Freeman and Co., San Francisco.

FISHER, J. C., (1974) *Energy Crises in Perspective,* John Wiley and Sons, New York.

MEADOWS, D. H., MEADOWS, D. L., RANDERS, J., BEHRENS III, W. W., (1972) *Limits to Growth,* Signet, New American Library, New York.

PRIEST, J., (1976) *Energy for a Technological Society,* Addison-Wesley, Reading, Mass.

DRYDEN, I. G. C., Ed., (1976) *The Efficient Use of Energy,* IPC Science and Technology Press, Guildford.

Chapter 2 Energy Resources

COMMITTEE ON RESOURCES AND MAN, NATIONAL ACADEMY OF SCIENCES AND NATIONAL RESEARCH COUNCIL, (1971) *Resources and Man,* W. H. Freeman and Co., San Francisco.

(1970) *Man's Impact on the Global Environment,* SCEP Report, MIT Press, Cambridge, Mass.

ROBINSON, N., Ed., (1966) *Solar Radiation,* Elsevier, Amsterdam.

RABINOWITCH, E., GOVINDJEE, (1969) *Photosynthesis,* John Wiley and Sons, New York.

BATEMAN, A. M., (1950) *Economic Mineral Deposits,* John Wiley and Sons, New York.

BORD NA MONA, (1972) *The Moving Bog,* Bord na Mona, Dublin.

LOWSON, M. H., Ed., (1970) Our Industry – Petroleum, British Petroleum Co. Ltd., London.

NATIONAL COAL BOARD, (1974) *Facts and Figures,* National Coal Board, London.

Chapter 3 Energy Conservation

(1974) *Turbine-generator Engineering,* GEC Turbine Generators Ltd., Manchester.
EDWARDS, J. D., (1973) *Electrical Machines,* International Textbooks Co. Ltd., Aylesbury.
BAKER, B. S., Ed., (1965) *Hydrocarbon Fuel Cell Technology,* Academic Press, New York.
BERKOWITZ, D. A., SQUIRES, A. M., Eds., (1971) *Power Generation and Environmental Change,* Chapters 10 and 11, MIT Press, Cambridge, Mass.
HUBBARD, M., (1967) *The Economics of Transporting Oil to and within Europe,* Maclaren, London.
BOCKRIS, J. O'M., (1976) *Energy: The Solar-Hydrogen Alternative,* Architectural Press, London.

Chapter 4 Power from Natural Sources

BRINKWORTH, B. J., (1972) *Solar Energy for Man,* Compton Press, Salisbury.
Solar Energy, Journal published for Solar Energy Society, Arizona State University, Tempe, Arizona, by Pergamon Press.
GUTHRIE-BROWN, J., Ed., (1970) *Hydroelectric Engineering Practice,* Vol. II, Blackie, London.
BERKOWITZ, D. A., SQUIRES, A. M., Eds., (1971) *Power Generation and Environmental Change,* Chapters 10 and 11, MIT Press, Cambridge, Mass.
PESKO, C., Ed., (1975) *Solar Directory,* Ann Arbor Sciénce Publication Inc.
GRAY, T. J., GASHUS, O. K., Eds., (1972) *Tidal Power,* Plenum Press, New York.

Chapter 5 Fossil Fuels

LOWSON, M. H., Ed., (1970) *Our Industry – Petroleum,* British Petroleum Co. Ltd., London.
BORD NA MONA, (1972) *The Moving Bog,* Bord na Mona, Dublin.
BRADBURY, K. A., (1973) *Solid Fuel in the Home,* Women's Solid Fuel Council, London.

Chapter 6 Nuclear Power

BENNET, D. J., (1972) *The Elements of Nuclear Power,* Longmans, London.
GLASSTONE, S., (1958) *Source Book on Atomic Energy,* D. van Nostrand Co. Inc., New York.
SALMON, A., (1964) *The Nuclear Reactor,* Longmans, London.

IAEA, (1975) *Le Phenomene D'Oklo,* IAEA, Vienna.

FLOWERS, B., (Chairman), (1976) *Sixth Report of the Royal Commission on Environmental Pollution,* HMSO, London.

Chapter 7 Inefficiencies

SKITT, J., (1972) *Disposal of Refuse and Other Waste,* Charter Knight and Co., London.

AMBROSE, E. R., (1966) *Heat Pumps and Electrical Heating,* John Wiley and Sons, New York.

DAVIES, S. J., (1950) *Heat Pumps and Thermal Compressors,* Constable, London.

DIAMANT, R. M. E., McGARRY, J., (1968) *Space and District Heating,* Iliffe, London.

Index